Arduino Electronics Blueprints

Make common electronic devices interact with an
Arduino board to build amazing out-of-the-box projects

Don Wilcher

BIRMINGHAM - MUMBAI

Arduino Electronics Blueprints

First published: July 2015

Production reference: 1200715

Published by Packt Publishing Ltd.
Livery Place
35 Livery Street
Birmingham B3 2PB, UK.

ISBN 978-1-78439-360-1

www.packtpub.com

Credits

Author
Don Wilcher

Reviewers
Samuel de Ancos

Liam Lacey

Commissioning Editor
Nadeem N. Bagban

Acquisition Editor
Harsha Bharwani

Content Development Editor
Ajinkya Paranjape

Technical Editor
Tejaswita Karvir

Copy Editor
Dipti Mankame

Project Coordinator
Harshal Ved

Proofreader
Safis Editing

Indexer
Priya Sane

Graphics
Sheetal Aute

Production Coordinator
Komal Ramchandani

Cover Work
Komal Ramchandani

About the Author

Don Wilcher is a passionate educator of electronics and robotics technology and an electrical engineer with 26 years of experience. He has authored several books on Lego Robotics and Arduinos. His latest book published by Maker Media, titled *Make: Basic Arduino Projects*, has been approved by the Alabama State Department of Education to be on their reading list. He's also a Certified Electronics Technician (CETa) and Exam Administrator (CA) with ETA International as well as State Certified Teacher for Career Technical Education (CTE) as a Level 2 Specialist in electronics technology. He has worked on industrial robotic systems, automotive electronic modules/systems, and embedded wireless controls for small consumer appliances. While at the Chrysler Corporation, Don developed a weekend enrichment pre-engineering program for inner city kids. He's a contributing writer and webinar lecturer for *Design News Magazine*. He is also an electronics and robotics technologist who has developed 21st century educational products and training materials for Makers, hardware start-up entrepreneurs, and educators.

I would like to thank my wife, Mattalene, and three children, Tiana, D'Vonn, and D'Mar, for their patience and understanding as I worked diligently to build and test Arduino circuits, coding, and writing during family time activities. This book is dedicated to you all. Also, I would like to thank my awesome Packt Publishing editors: Ajinkya Paranjape (content development editor), Harsha, Bharwani (acquisition editor), and Tejaswita V. Kavir (technical editor) for your patience, dedication, comments, and great suggestions on creating a wonderful book. I look forward to working with you all soon on another book project.

About the Reviewers

Samuel de Ancos lives in Madrid, Spain. He loves developing software. He is currently working as a senior engineer at Carriots.com IoT and M2M platform and is a member of the Fourcoders software developers team. He writes a blog about software developing (in Spanish).

He has more than 7 years of experience in developing web applications with PHP / MySql / MongoDB using frameworks such as Symfony 1.4 / 2.x / Silex and also has more than 5 years of experience with Python, knowledge of the Tornado Web Server framework, Web.py framework, and the Bottle framework. He has more than 4 years of experience in developing the IoT and M2M platforms with knowledge of MongoDB, RabbitMQ, and Redis.

Liam Lacey is a software developer who specializes in C/C++ development, mainly in the fields of audio and MIDI, of OS-level applications. Most of his development skills have been self-taught through a strong passion for the field; however, he was first introduced to software development at the university, where he received a first class BSc honors in audio and music technology. He has designed and developed software modules for new products, from the concept/prototype stage all the way through to the production stage, within teams and as an independent developer.

He has a strong interest in audio plugin development, connected devices, music interaction, new interfaces for musical expression, and augmented instruments using platforms such as Arduino and JUCE. He is also a guitarist, musician, composer, producer, performer, and sound designer.

He aims to develop his current software development skills toward a highly professional level as well as develop skills in other related areas, such as audio DSP, sound synthesis and design, and electronics.

www.PacktPub.com

Support files, eBooks, discount offers, and more

For support files and downloads related to your book, please visit www.PacktPub.com.

Did you know that Packt offers eBook versions of every book published, with PDF and ePub files available? You can upgrade to the eBook version at www.PacktPub.com and as a print book customer, you are entitled to a discount on the eBook copy. Get in touch with us at service@packtpub.com for more details.

At www.PacktPub.com, you can also read a collection of free technical articles, sign up for a range of free newsletters and receive exclusive discounts and offers on Packt books and eBooks.

https://www2.packtpub.com/books/subscription/packtlib

Do you need instant solutions to your IT questions? PacktLib is Packt's online digital book library. Here, you can search, access, and read Packt's entire library of books.

Why subscribe?

- Fully searchable across every book published by Packt
- Copy and paste, print, and bookmark content
- On demand and accessible via a web browser

Free access for Packt account holders

If you have an account with Packt at www.PacktPub.com, you can use this to access PacktLib today and view 9 entirely free books. Simply use your login credentials for immediate access.

Table of Contents

Preface

You have purchased your first Arduino, and now you're wondering what project to build with it. There are hundreds of websites with an assortment of electronic gadgets and devices to build, but the search and choosing the first project can be overwhelming. Besides building awesome Arduino gadgets, some of the website projects leave out how the electronics and code work with a programmable prototyping platform. Also, the projects found on the Web don't provide additional challenges to test your new Maker skills as well.

The *Arduino Electronics Blueprints* book was written to address the concerns mentioned in a user friendly and educational format. Every chapter in the book starts off with either a historical reference to electronic discoveries or a brief discussion of present technologies used in contemporary consumer, entertainment, or industrial products. The book was designed to show how to build awesome electronic devices using parts found in laboratory bins or junk boxes. Also, new prototyping materials such as littleBits electronic modules and Elenco SNAP circuit kits are introduced to readers as well. The new and exciting prototyping materials presented allow us to rapidly build the target Arduino device discussed in some of the book's chapters. To aid readers in building the fun Arduino projects, a Parts list of electronic components is included in each chapter of the book. Detailed circuit schematic and wiring diagrams and Arduino code are provided in each chapter. Also, basic circuit theory and Arduino code explanations are provided in each project chapter as well. To conclude the chapter, a DIY challenge is presented, so readers may explore additional prototyping topics in new product designs of their own. I enjoyed designing, building, and testing each chapter's project and hope readers of the *Arduino Electronics Blueprints* book will find the projects to be fun and entertaining as well.

What this book covers

Chapter 1, A Sound Effects Machine, will teach the reader how to build an Arduino sound effects machine using an SD module, digital logic switches, a transistor speaker amplifier, and .wav files. Also, the reader will learn how to add a random function in order as to play different sounds automatically without using digital logic switches.

Chapter 2, Programmable DC Motor Controller with an LCD, shows the reader how to build an Arduino programmable controller to operate small DC motors. Also, to aid in operating the programmable controller, the reader will learn how to add a Liquid Crystal Display (LCD) to the electronic device as well.

Chapter 3, A Talking Logic Probe, explains a talking electronic instrument that the reader can build to test microcontroller and digital circuits. Also, the reader will learn how to wire an Arduino to an EMIC 2 (text-to-speech) module and program it using special character codes.

Chapter 4, Human Machine Interface, A Human Machine Interface (HMI) used in industrial controls to operate electromechancial devices, such as motors, will be discussed in this chapter. The reader will also learn how to build a HMI using an Arduino, a virtual server, and JavaScript to control a small DC motor.

Chapter 5, IR Remote Control Tester, allows the reader to learn how to build a testing device to check the operation of any IR remote control. Also, the reader will learn about IR detectors and digital codes using this electronic tester.

Chapter 6, A Simple Chat Device with LCD, will teach the reader how to send text messages to an Arduino using a Bluetooth Low Energy (BLE) device and an Android smartphone. Also, the RedBearLabs BLE Arduino shield used to send and receive text messages will be introduced to the reader in this chapter.

Chapter 7, Bluetooth Low Energy Controller, will show the reader how to send BLE control signals to an Arduino using the RedBearLabs BLE shield and an Android smartphone to control a DC motor. Also, a seven segment LED display's electrical operation will be discussed along with making letter characters using the BLE Controller.

Chapter 8, Capacitive Touch Sensing, explores a simple DC motor controller using an Arduino and a 555 timer IC-based capacitive touch sensor. The reader will learn the basic operation of the 555 timer by building an Arduino-enabled touch sensing controller.

Chapter 9, Arduino-SNAP Circuit AM Radio, introduces the reader the Elenco SNAP circuit kit by building an AM radio. Also, the reader will learn how to operate the AM radio using the RedBearLab BLE shield and an Android smartphone.

Chapter 10, Arduino Scrolling Marquee, discusses organic light-emitting diode (OLED) technology by building an Arduino-based scrolling marquee. Also, the reader will learn to use any ordinary IR handheld remote to control the scrolling effect of the OLED marquee.

What you need for this book

To build the awesome electronics gadgets and devices in this book, the following materials are required:

- An Arduino Uno (Rev 3 electronics board).
- The latest Arduino IDE can be downloaded from the website `http://www.arduino.cc/en/Main/Software`.
- An assortment of electronic components (resistors, capacitors, transistors, diodes, seven segment LED display, 74LS04 Hex inverter IC, and 74LS00 NAND logic gate IC).
- The EMIC 2 text-to-speech module. The module can be purchased from Parallax Inc's website `https://www.parallax.com/product/30016`.
- littleBits deluxe set. The electronics module kit can be purchased from littleBits website `http://littlebits.cc/shop`.
- The Elenco SNAP circuit kit. The Elenco SNAP circuit kit can be purchased from Adafruit website `https://www.adafruit.com/category/117`.
- A solderless breadboard.
- A jumper wire kit.
- A small variable DC power supply (the variable output voltage rating of 0-24V DC with an output current rating of 2A max).
- 9V batteries with battery snap connectors.

Who this book is for

This book is intended for those who want to learn about electronics and coding by building amazing devices and gadgets with the Arduino. If you are an experienced developer who understands the basics of electronics, then you can quickly learn how to build smart devices using the Arduino. Perhaps you have never used electronic components and are new to the Arduino, but have coding skills. In either case, this book will provide you with the knowledge to build amazing, smart, and fun-to-use devices. The only experience needed is a desire to learn about electronics, circuit breadboarding, and coding.

Conventions

In this book, you will find a number of text styles that distinguish between different kinds of information. Here are some examples of these styles and an explanation of their meaning.

Code words in text, database table names, folder names, filenames, file extensions, pathnames, dummy URLs, user input, and Twitter handles are shown as follows: "The `<Serial.h>` library allows the text message to be converted into its equivalent ASCII code."

A block of code is set as follows:

```
void loop(){

  // read the status of the Program Switch value:
  ProgramStatus = digitalRead(ProgramPin);

  // check if Program switch is ON.
  if(ProgramStatus == HIGH) {
    digitalWrite(OUTPin, HIGH);

  }
  else{
      digitalWrite(OUTPin,LOW);

  }
}
```

New terms and **important words** are shown in bold. Words that you see on the screen, for example, in menus or dialog boxes, appear in the text like this: "Click on the **Connect** button on the Breakout server application."

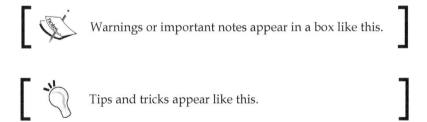

Warnings or important notes appear in a box like this.

Tips and tricks appear like this.

Reader feedback

Feedback from our readers is always welcome. Let us know what you think about this book—what you liked or disliked. Reader feedback is important for us as it helps us develop titles that you will really get the most out of.

To send us general feedback, simply e-mail feedback@packtpub.com, and mention the book's title in the subject of your message.

If there is a topic that you have expertise in and you are interested in either writing or contributing to a book, see our author guide at www.packtpub.com/authors.

Customer support

Now that you are the proud owner of a Packt book, we have a number of things to help you to get the most from your purchase.

Downloading the example code

You can download the example code files from your account at http://www.packtpub.com for all the Packt Publishing books you have purchased. If you purchased this book elsewhere, you can visit http://www.packtpub.com/support and register to have the files e-mailed directly to you.

Downloading the color images of this book

We also provide you with a PDF file that has color images of the screenshots/diagrams used in this book. The color images will help you better understand the changes in the output. You can download this file from: http://www.packtpub.com/sites/default/files/downloads/3601OS_ColoredImages.pdf.

Errata

Although we have taken every care to ensure the accuracy of our content, mistakes do happen. If you find a mistake in one of our books—maybe a mistake in the text or the code—we would be grateful if you could report this to us. By doing so, you can save other readers from frustration and help us improve subsequent versions of this book. If you find any errata, please report them by visiting http://www.packtpub.com/submit-errata, selecting your book, clicking on the **Errata Submission Form** link, and entering the details of your errata. Once your errata are verified, your submission will be accepted and the errata will be uploaded to our website or added to any list of existing errata under the Errata section of that title.

To view the previously submitted errata, go to `https://www.packtpub.com/books/content/support` and enter the name of the book in the search field. The required information will appear under the **Errata** section.

Piracy

Piracy of copyrighted material on the Internet is an ongoing problem across all media. At Packt, we take the protection of our copyright and licenses very seriously. If you come across any illegal copies of our works in any form on the Internet, please provide us with the location address or website name immediately so that we can pursue a remedy.

Please contact us at `copyright@packtpub.com` with a link to the suspected pirated material.

We appreciate your help in protecting our authors and our ability to bring you valuable content.

Questions

If you have a problem with any aspect of this book, you can contact us at `questions@packtpub.com`, and we will do our best to address the problem.

1
A Sound Effects Machine

Arduino is a wonderful rapid prototyping platform capable of making a variety of electronic tools, gadgets, and instruments. The fascination with Arduino has grown to the point where makers are creating devices to educate, entertain, and provide new creative tools for individuals interested in science or technology. Some of the electronic devices that can be built with Arduino include DC motor controllers, musical instruments, robots, and smart toys.

In addition to the devices listed, Arduino can be used to create unique sounds as well. There are several programming techniques used to create sounds using Arduino. For example, the Arduino `tone` library can be used to create a variety of unusual sounds by varying the pitch, frequency, and duration of a pulse width modulated signal. Another approach is to use the recorded sound WAV files stored on an SD card. Arduino can select these files using an SD card software library, thereby allowing the recorded sounds to be heard through a small signal transistor amplifier. In this chapter, we will explore how to build a sound effects machine capable of playing a variety of sounds recorded on an SD card. Also, a discussion on small signal transistor amplifiers and **Serial Peripheral Interface** (**SPI**) communication will be reviewed with the project. A modification project allowing the sounds to be played randomly will be discussed in this chapter as well.

Parts list
The following is the list of parts required for a sound effects machine:

- Arduino Uno (one unit)
- 2N3904 NPN transistor (one unit)
- 8 ohm speaker (one unit)

- 470 ohm resistor (yellow, violet, brown, and gold) (one unit)
- 4.7 kilo ohm resistors (yellow, violet, red, and gold) (four units)
- Tactile pushbutton or toggle switches (four units)
- SD module (DFRobot [PN:DFR0071]) or Arduino compatible (one unit)
- Breadboard
- Wires

A sound effects machine block diagram

Building a sound effects machine is relatively easy as this device only requires four electronic subcircuits, as shown in the following block diagram:

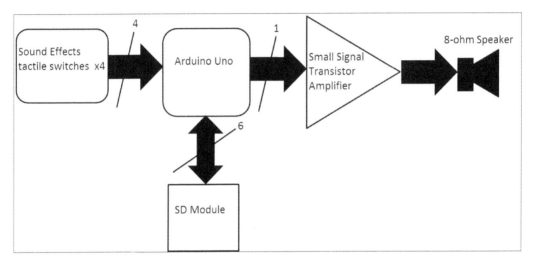

The blocks in the diagram represent the electronic subcircuits.The arrows are the electrical wires connecting to the target electronic subcircuit. The numbers shown with the diagonal lines attached to the arrows tell the number of wires connected to each electronic subcircuit.

The sound effects machine block diagram is an engineering tool used to convey a complete product design using simple graphics. The block diagram also makes it easier to plan the breadboard for prototyping and testing of the sound effects machine in a workshop or laboratory bench. One final observation of the sound effects machine block diagram is that the basic computer convention of inputs is on the left, the processor is located in the middle, and the outputs are placed on the right-hand side of the design layout. As shown, the tactile switches are on the left-hand side, the Arduino is located in the middle, and the small signal transistor amplifier with 8 ohm speaker is on the right-hand side of the block diagram. This left to right design method makes it easier to build the sound effects machine and troubleshoot the errors during the testing phase of project development.

Building the sound effects machine

The sound effects machine is quite simple in design and construction. There are a variety of ways to build this electronic device, such as on a **Printed Circuit Board (PCB)** or an experimenter board. The method I found to rapidly build this electronic device is to use a solderless breadboard, as shown in this diagram:

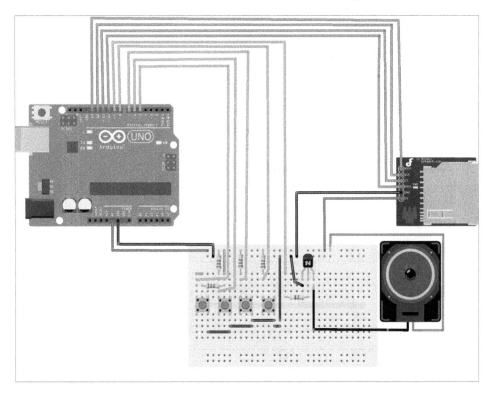

In the preceding wiring diagram, the electrical components to build the sound effects machine circuit are placed on the solderless breadboard for easy attachment to the Arduino, SD card module, and speaker. The transistor shown is a 2N3904 NPN type with pin-out arrangement consisting of emitter, base, and collector, in that order. If the transistor pins are wired incorrectly, the WAV file sounds will not be heard through the 8 ohm speaker. The SD module is a DFRobot component (shown next) but any compatible part can be used in this project. If a compatible SD module component is being used, wire the device to the Arduino using the manufacturer's datasheet to ensure proper circuit operation.

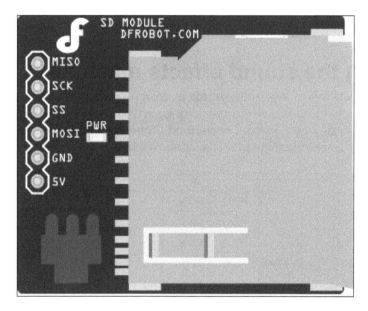

Another wiring tool to be used in building the sound effects machine, either on a solderless breadboard or an experimenter board, is a circuit schematic diagram shown next:

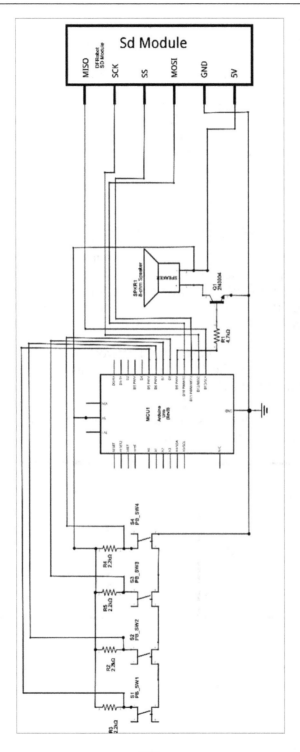

As illustrated, the tactile pushbutton switches are used to select the various sounds stored on the SD card. Along with the tactile switches, there are four 2.2 kilo ohm pull-up resistors wired to them. The pull-up resistors allow a binary 0 to be read by the Arduino for selecting the stored WAV files on the SD card.

There are a total of five sound files stored on the SD card. Four of these sound files are easily activated by pressing a tactile pushbutton switch. The fifth sound file turns on when the Arduino is powered on or manually reset. To play the stored sounds on the SD card, the Arduino digital pins D5-D8 require a binary 0 to select the WAV files. More sounds can be added to the SD card and activated by wiring additional tactile pushbutton switches with 2.2 kilo ohm pull-up resistors.

Introducing SPI communication

The SPI communication is used to connect electronic devices such as computers, scanners, and printers together with four wires. The communication concept behind SPI is based on a master-slave configuration. The communication between the master and slave units are full duplex. In full duplex mode, the signals are transmitted simultaneously between the master and slave devices.

An example of a full duplex communication network is a desktop computer connected to a laser printer. The desktop computer will send a message to the laser printer, alerting it that a document is ready for printing. The laser printer will acknowledge the desktop computer by sending a request to print message. Upon receiving the request to print message, the desktop computer will then send the document to the laser printer for printing.

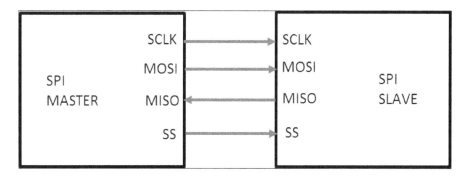

Inside the master and slave units, there are 8-bit shift registers that send binary data between them. A digital clock is used to provide proper timing to the shift registers while data moves between the two connected devices. The binary data movement used by both the shift registers is a *Shift Left-Rotate Right* sequence.

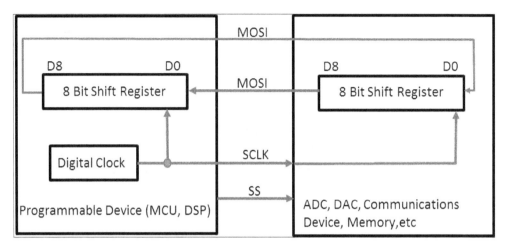

The data line names for the SPI 4-wire communication network are:

- MOSI (Master Out, Slave In)
- MISO (Master In, Slave Out)
- SCLK (Serial Clock)
- SS (Slave Select)

Most slave Integrated Circuit(IC) manufacturers (for example, DFRobot) use different pin names for the SPI 4-wire communication configuration. Some of these pin names are listed here:

- DI (Digital-In)
- DO (Digital-Out)
- SCK (Serial Clock)
- CS (Chip Select)

Additional slaves can be attached to the master SPI device by wiring them in parallel. The slave select signal will be connected from the master to each of the slaves. The slave IC's output is enabled when a corresponding slave signal is turned *on* or *set*. The data output from the slave IC is disconnected from the MISO line when the device is deselected by the slave select line.

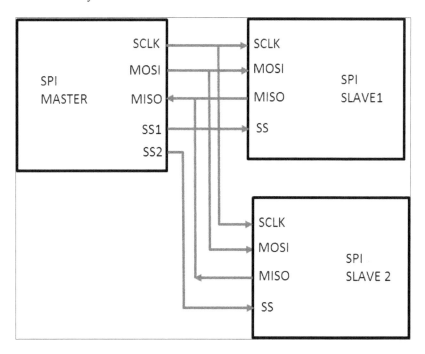

The wiring for the Arduino to the SD module uses the techniques just discussed to select the individual files on the card. The circuit schematic diagram provides another way to build the microcontroller to SD module interface for sound WAV file selection:

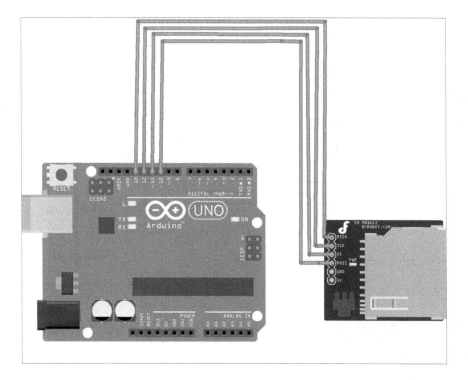

A circuit schematic showing the electrical wiring connections between the Arduino Uno and SD module is shown here:

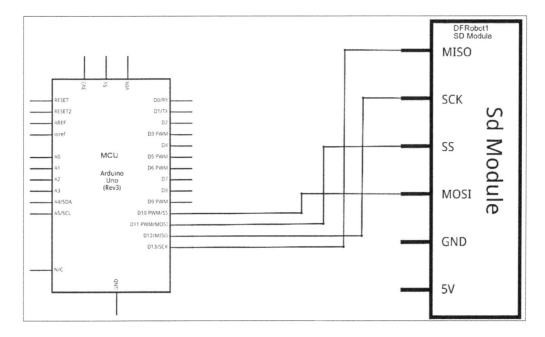

Adding digital logic switches for WAV file selection

The sound effects machine can produce a multitude of strange noises stored on an SD card. The eerie sounds can be recorded using a free audio editor and recorder called **Audacity**. Audacity can be found on the Web at http://soundbible.com/tags-strange.html.

Once the desired files have been recorded and saved on the SD card, accessing them requires using digital logic switches wired to the Arduino, as shown here:

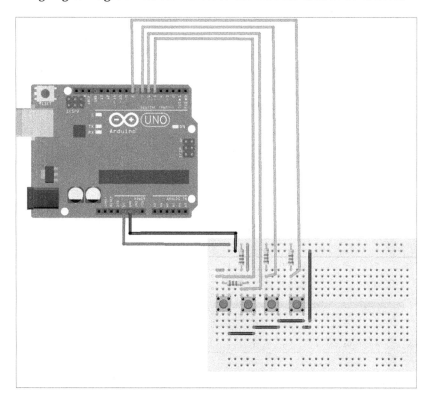

The four tactile pushbutton switches along with the 2.7 kilo ohm resistors are wired as Active Low digital inputs. An Active Low operation is a binary 0 value that exists when a trigger, such as a pushbutton switch, has been initiated. The Arduino digital inputs D5-D8 are wired to the four tactile pushbutton switches for receiving the binary 0 digital data. The Arduino sketch will select the appropriate WAV file based on its pushbutton switch being pressed. The following is a small line of the Arduino code used to read an Active Low trigger from a pushbutton tactile switch:

```
if (SW1 == LOW) { //if SW1 pressed then play file "6.wav"
    tmrpcm.play("6.wav");
}
```

The line of code will play the WAV file 6.wav only when the pushbutton tactile switch is pressed, triggering an Active Low input signal to the Arduino.

Adding SD and WAV file libraries to your Arduino sketch

To allow the sound effects machine to play WAV files, additional software resources are needed. The two important software resources required for the sound effects machine to work properly are the `<SD.h>` and `<TMRpcm.h>` libraries. The **SD (San-Disk)** library is required for read / write operations from an SD part. The **Arduino Integrated Development Environment (IDE)** has the SD resource as part of its software library.

With the SD library included in the Arduino sketch, as shown next, the available software resource allows the microcontroller to talk to the actual SD card using a module or breakout board. The pin allowing the Arduino to talk to the SD card module or breakout board is the SS or CS pin:

```
#include <SD.h>    // need to include the SD library

#define SD_ChipSelectPin 10  //using digital pin 10 on Arduino 328

if (!SD.begin(SD_ChipSelectPin)) {  // see if the card is present
  and can be initialized:
  return;   // don't do anything more if not
}
```

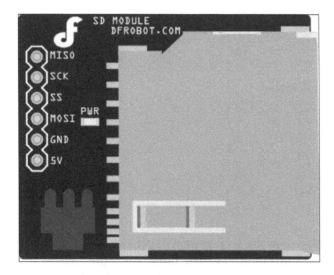

Depending on the SD module or breakout board manufacturer, the SS or CS pin assignment on the Arduino may vary. Therefore, consult the manufacturer's datasheet for the appropriate Arduino pin assignment. With the appropriate Arduino pin numbers assigned, another file to include in the Arduino sketch is the TMRpcm library.

The TMRpcm library installation

The TMRpcm is a software library used for playing PCM/WAV files on Arduino. The Arduino SD library is used to communicate with an SD card to read data, such as WAV files. The WAV files can be accessed using the TMRpcm library so that they may be heard through a speaker wired to the Arduino. Perform the following steps to install the TMRpcm library:

1. To obtain the TMRpcm library to use it in the sound effects machine project, there is a **GitHub** website—https://github.com/TMRh20/TMRpcm/wiki.

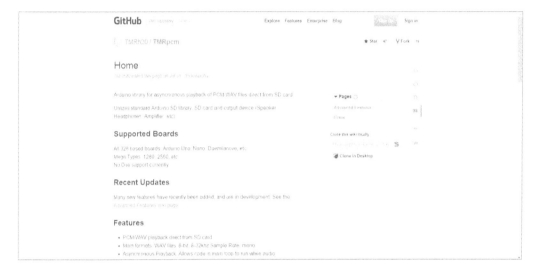

2. Scrolling down on the web page will provide a link to the Arduino.cc website for the library installation instructions. Also, the TMRpcm zipped file is located above this link and can be downloaded on your desktop, pc, or notebook computer's hard drive.

3. After the installation has been completed, the TMRpcm file will be included in the Arduino's IDE library.

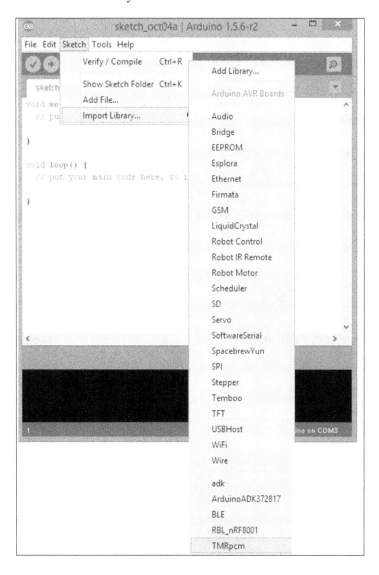

4. With the TMRpcm file included in the Arduino IDE library, the following line of code can be added to the sound effects machine sketch:

```
#include <SD.h>              // need to include the SD library
#define SD_ChipSelectPin 10  //using digital pin 4 on
Arduino nano 328
#include <TMRpcm.h>           //  also need to include this
library...
#include <SPI.h>
```

5. Also included with the sketch is the <SPI.h> file for connecting the SD module or breakout board to the Arduino. With the TMRpcm library added to the Arduino sketch, the WAV file audio content can be heard using a simple transistor amplifier, as shown in the following image:

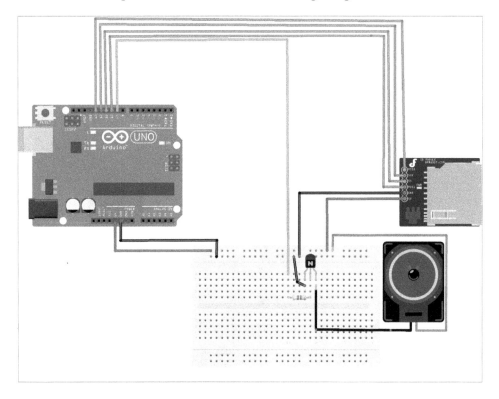

6. The circuit schematic can be used to wire the transistor amplifier to the Arduino, as shown in the next diagram:

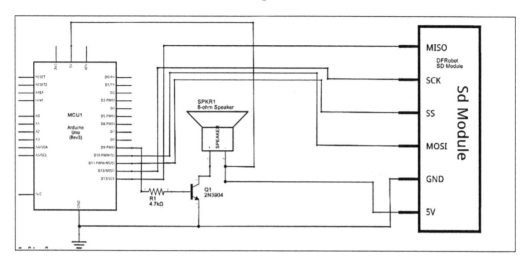

The 2N3904 NPN transistor has sufficient gain or volume potential (hfe 100 typical) to drive an 8 ohm speaker. The hfe is an electrical symbol parameter found on transistor or audio amplifier integrated circuit datasheets. The unitless number refers to the current gain for audio amplifiers and transistors to sufficiently drive an electromechanical load wired to them. For amplifiers, in particular, sometimes it refers to the amplification factor of increasing an input signal applied to the transistor or IC amplifier. If a higher audio volume is desired, an LM386 power amplifier IC can easily be substituted for the simple transistor circuit.

7. With the `<TMRpcm.h>` library included in the Arduino sketch, the following lines of code allow the WAV files to be heard through the speaker driven by the transistor amplifier:

```
TMRpcm tmrpcm;    // create an object for use in this sketch
   int SW1;
   int SW2;
   int SW3;
   int SW4;
   void setup(){
```

```
      pinMode(5,INPUT);   //Define A0 as digital input.
      pinMode(6,INPUT);   //Define A1 as digital input.
      pinMode(7,INPUT);   //Define A2 as digital input.
      pinMode(8,INPUT);   //Define A3 as digital input.

      tmrpcm.speakerPin = 9; //11 on Mega, 9 on Uno, Nano,
      etc

      if (!SD.begin(SD_ChipSelectPin)) {
        // see if the card is present and can be initialized:
        return;    // don't do anything more if not
      }
      tmrpcm.volume(7);
      tmrpcm.play("1.wav"); //the sound file "1" will play
      each time the Arduino powers up, or is reset
    }

    void loop(){
      SW1=digitalRead(5);
      SW2=digitalRead(6);
      SW3=digitalRead(7);
      SW4=digitalRead(8);

      if (SW1 == LOW) { //if SW1 pressed then play file
        "6.wav"
        tmrpcm.play("6.wav");
      } else if(SW2 == LOW){ //if SW2 pressed then play file
        "4.wav"
        tmrpcm.play("4.wav");
      } else if(SW3 == LOW){ //if SW3 pressed then play file
        "5.wav"
        tmrpcm.play("5.wav");
      } else if(SW4 == LOW){  //if SW4 pressed then play file
        "3.wav"
        tmrpcm.play("3.wav");
      }

    }
```

As you can see, this code follows the traditional coding format for programming an Arduino. After uploading the sketch to the Arduino, 1.wav will be heard through the speaker for a few seconds. This initial tone heard upon resetting the Arduino alerts you that the sound effects machine is ready to use. By pressing each of the pushbutton tactile switches, a unique sound will be heard through speaker. If an automatic approach is desired, a random function can easily be programmed for the sound effects machine.

Adding a random function to play sounds automatically

To allow the sound effects machine to select various sounds automatically, the random function can be used. The description for the random function is given on the Arduino.cc website:

> Description
>
> The random function generates pseudo-random numbers.

The key concept behind this programming instruction is to set a maximum number for the random function generator. This maximum number will prevent the random function from generating values not exceeding the set maximum limit. For example, random(300) will allow the generation of numbers ranging from 0 to 300 to be produced in a non-sequential matter. With this function, the sound effects machine will select WAV files automatically to play, and they can be heard through the simple transistor amplifier-speaker circuit.

To test this random generator function, type the following sketch onto the text editor of the Arduino IDE:

```
long randNumber;
void setup(){
  Serial.begin(9600);
}

void loop(){
  randNumber = random(5); // set max random number
  Serial.println(randNumber); //print random number on Serial
Monitor
  delay(1000); //wait 1 sec between printing random numbers
}
```

Attach an Arduino to the desktop pc or notebook computer and upload the sketch to it. Next, open the serial monitor and notice the numbers being displayed on the screen. The numbers generated will range between 0 to 5 on the serial monitor. Modeling this function on the Arduino IDE allows a better understanding of how the concept of selecting sounds from the SD card randomly is feasible. Try out different values for the random function parameter and watch the results on the serial monitor.

The next phase of the sound effects machine modification project will consist of taking sections of the random function and adding them to the original sketch. The technique used in this modification method is called **software remixing**. This software remixing method allows Proof of Concept-embedded devices to be rapidly developed and tested. Again, our concept is to provide a hands-free sound effects machine for selecting WAV files from an SD card.

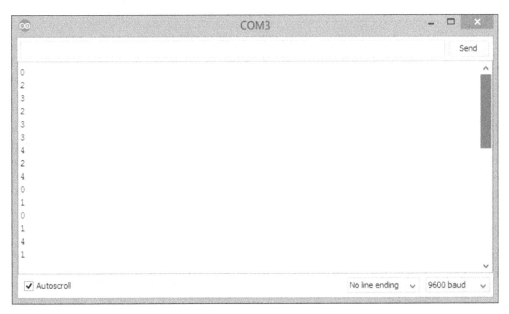

Type the following sketch onto the IDE text editor and upload the code to the Arduino:

```
#include <SD.h>                    // need to include the SD library
#define SD_ChipSelectPin 11b  //using digital pin 4 on Arduino nano
328
#include <TMRpcm.h>             //  also need to include this
library...
#include <SPI.h>
long randNumber;

TMRpcm audio;    // create an object for use in this sketch

void setup(){
   audio.speakerPin = 9; //11 on Mega, 9 on Uno, Nano, etc
```

```
  if (!SD.begin(SD_ChipSelectPin)) {  // see if the card is
    present and can be initialized:
    return;    // don't do anything more if not
  }
  audio.volume(7);
  audio.play("1.wav"); //the sound file "1" will play each time
  the Arduino powers up, or is reset
}

void loop(){
  randNumber = random(5); // set max random number

  if (randNumber == 5){ // if number is 5 play file ""6.wav"
    audio.play("6.wav");
    delay(5000);

  } else if(randNumber == 4){ // if number is 4 play file "4.wav"
      audio.play("4.wav");
      delay(5000);

  } else if(randNumber == 3){ // if number is 3 play file "5.wav"
    audio.play("5.wav");
    delay(5000);

  } else if(randNumber == 2){ // if number is 2 play file "3.wav"
    audio.play("3.wav");
    delay(5000);

  } else if (randNumber == 0){ // if number is 0 play "1.wav"
    audio.play("1.wav");
    delay(5000);
  } else if (randNumber == 1){ // if number is 1 play "2.wav"
    audio.play("2.wav");
    delay(5000);
  }


}
```

The WAV files will play randomly, based on the number stored in the randNumber variable. The delay(5000) instruction will allow the selected sound to play for 5 sec. An added feature of this sketch is that new files can be added to the SD card and can also be selected on the basis of the assigned number chosen by the else if statement. If additional sounds are added with the random (n) instruction, they will be adjusted to accommodate the number of new WAV files stored on the SD card. Also, to see which WAV files have been selected, the serial monitor can be modified to include the new lines of code.

In addition, the actual WAV file name can be displayed as the machine plays the SD card audio content using the serial monitor. A few additional lines of code can be included to the sketch shown earlier. The new changes to the Random Function WAV code are given here:

```
#include <SD.h>                    // need to include the SD library
#define SD_ChipSelectPin 11  //using digital pin 4 on Arduino nano
328
#include <TMRpcm.h>            //  also need to include this
library...
#include <SPI.h>
long randNumber;

TMRpcm audio;    // create an object for use in this sketch

void setup(){
  Serial.begin(9600);
  audio.speakerPin = 9; //11 on Mega, 9 on Uno, Nano, etc

  if (!SD.begin(SD_ChipSelectPin)) {  // see if the card is
    present and can be initialized:
    return;    // don't do anything more if not
  }
  audio.volume(7);
  audio.play("1.wav"); //the sound file "1" will play each time
  the Arduino powers up, or is reset
}

void loop(){
  randNumber = random(5); // set max random number

  if (randNumber == 5){ // if number is 5 play file "6.wav"
```

```
    audio.play("6.wav");
    Serial.println("Playing 6.wav");
    delay(5000);

} else if(randNumber == 4){ // if number is 4 play file "4.wav"
    audio.play("4.wav");
    Serial.println("Playing 4.wav");
    delay(5000);

} else if(randNumber == 3){ // if number is 3 play file "5.wav"
    audio.play("5.wav");
    Serial.println("Playing 5.wav");
    delay(5000);

} else if(randNumber == 2){ // if number is 2 play file "3.wav"
    audio.play("3.wav");
    Serial.println("Playing 3.wav");
    delay(5000);

} else if (randNumber == 0){ // if number is 0 play "1.wav"
    audio.play("1.wav");
    Serial.println("Playing 1.wav");
    delay(5000);
} else if (randNumber == 1){ // if number is 1 play "2.wav"
    audio.play("2.wav");
    Serial.println("Playing 2.wav");
    delay(5000);
}

}
```

Downloading the example code

You can download the example code files from your account at
http://www. packtpub. com for all the Packt Publishing books
you have purchased. If you purchased this book elsewhere, you
can visit http://www.packtpub.com/support and register to
have the files e-mailed directly to you.

Type the new sketch onto the text editor of the Arduino's IDE. Upload the sketch to the USB connected Arduino and open the serial monitor. As sounds are playing through the speaker, the WAV file names will be displayed on the serial monitor:

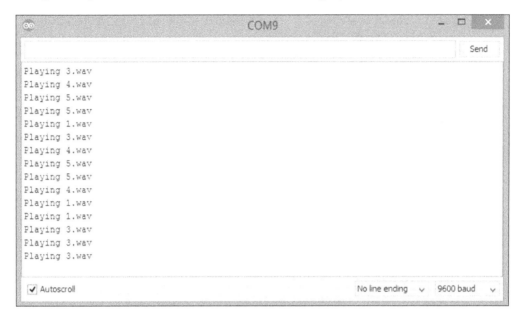

The serial monitor provides a user friendly visual display to watch the WAV sound files as they are being played. Also, this monitor can serve as a software debugging tool by providing WAV file data as the code is being executed. If a particular file is not providing sound, the serial monitor will validate the absence of the data on the SD card quite easily.

Adding an LED bar graph display for selected sound

Besides displaying the names of the sound files on a serial monitor, a lighting indicator can be used for positive visual feedback. The idea behind this new feature is to provide a unique visual effect for the random function. The random patterns generated within the software can be seen using solid state indicators like LEDs. The visual random patterns using LEDs will look best when viewed in a darkroom. The **Proof of Concept (POC)** that is being presented in this section is more of a design challenge to take the knowledge obtained from the previous hardware and software changes for building a unique and visually appealing device.

To help develop the LED bar graph display POC, a possible breadboard layout is shown in the following figure:

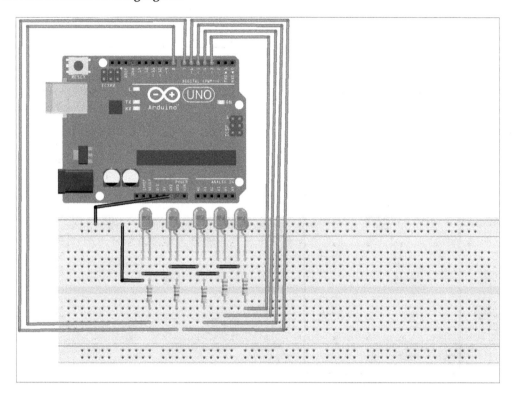

As shown in the breadboard, five single LEDs along with 330 ohm series resistors are used to monitor the associate sound files stored on the SD card. The idea behind the breadboard wiring is to show that digital output pins are needed to operate (drive) the LEDs. The digital pins *D3*, *D4*, *D5*, *D7*, and *D8* are used as output drivers for operating the five LEDs.

A circuit schematic diagram shows a better wiring perspective of the LEDs, series resistors, and their attachment to the Arduino:

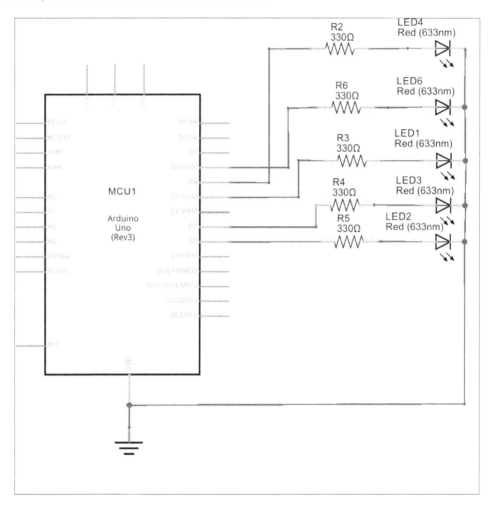

Although the wiring diagram shows red LEDs, other colors maybe used for the random function visual effect of the code. Also, the five LEDs can be replaced with an LED bar graph display, such as Sparkfun Electronics' catalog number COM-09937. Although the LED bar graph display has 10 individual LEDs in one package, only five are required for this project.

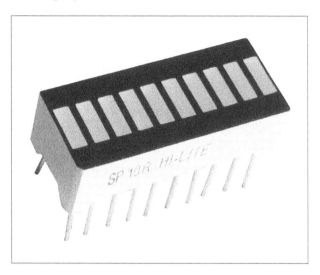

Once the circuit breadboard wiring is complete, a new code is required to operate the LED bar graph display. The following code blocks are required for the Random Function WAV sketch. These example code blocks are templates to help modify the sound effects machine for visual display of the sounds randomly being selected by the code. The digital pins shown in the code will need to be declared, as shown in the code block; the constant names chosen reflect the optoelectronic component wired to the designated digital pins on the Arduino:

```
int LEDBar1 = 3; // segment 1 of the LED Bar Graph display
int LEDBar2 = 4; // segment 2 of the LED Bar Graph display
int LEDBar3 = 5; // segment 3 of the LED Bar Graph display
int LEDBar4 = 7; // segment 4 of the LED Bar Graph display
int LEDBar5 = 8; // segment 5 of the LED Bar Graph display
```

Note that the declaration of constant names and variables are placed above the `void` `setup()` section of the code. Next, the constant names need to be configured as digital outputs for the Arduino:

```
void setup(){
  pinMode(LEDBar1,OUTPUT);  //Define LEDBar1 as digital output.
  pinMode(LEDBar2,OUTPUT);  //Define LEDBar2 as digital output
  pinMode(LEDBar3,OUTPUT);  //Define LEDBar3 as digital output
  pinMode(LEDBar4,OUTPUT);  //Define LEDBar4 as digital output
  pinMode(LEDBar5,OUTPUT);  // Define LEDBar5 as digital output
}
```

The final step of adding the LED bar graph display to the random function WAV code is to operate the LEDs when the assigned sound file is selected:

```
void loop(){
  randNumber = random(5); // set max random number

  if (randNumber == 5){ // if number is 5 play file "6.wav"
    audio.play("6.wav");
    Serial.println("Playing 6.wav");
    digitalWrite(LEDBar1, HIGH);
    delay(5000);

  } else if(randNumber == 4){ // if number is 4 play file "4.wav"
    audio.play("4.wav");
    Serial.println("Playing 4.wav");
    digitalWrite(LEDBar1, LOW);
    digitalWrite(LEDBar2, HIGH);
    delay(5000);

  } else if(randNumber == 3){ // if number is 3 play file "5.wav"
    audio.play("5.wav");
    Serial.println("Playing 5.wav");
    digitalWrite(LEDBar2, LOW);
    digitalWrite(LEDBar3, HIGH);
    delay(5000);

  } else if(randNumber == 2){ // if number is 2 play file "3.wav"
    audio.play("3.wav");
    Serial.println("Playing 3.wav");
    digitalWrite(LEDBar3, LOW);
```

```
    digitalWrite(LEDBar4, HIGH);
    delay(5000);

} else if (randNumber == 0){ // if number is 0 play "1.wav"
    audio.play("1.wav");
    Serial.println("Playing 1.wav");
    digitalWrite(LEDBar4, LOW);
    digitalWrite(LEDBar5, HIGH);
    delay(5000);

} else if (randNumber == 1){ // if number is 1 play "2.wav"
    audio.play("2.wav");
    Serial.println("Playing 2.wav");
    digitalWrite(LEDBar5, LOW);
    delay(5000);
    }

}
```

With these suggested changes to the Random Function WAV code, the sound effects machine will have a visually appealing look as WAV files are being played randomly.

Summary

In this chapter, a sound effects machine was built using an Arduino, SD module, and few off-the-shelf electronic components. The fundamentals of SPI and full duplex communications were explained in the chapter. Adding digital logic switches to the Arduino and WAV files to the SD card were also explained. The installation of the TMRpcm library was provided using GitHub website images and step by step instructions. The operation of the Arduino code was explained aided by the comment statements in the applications software. Finally, the sound effects machine was tested along with the additional information to provide a random file selection feature. Adding an LED bar graph display for selected sounds was explained.

In *Chapter 2, Programmable DC Motor Controller with LCD*, a discussion on how to build a small electronic controller to operate small power, low voltage DC rotating machines will be presented. Learning objectives in the chapter include interfacing discrete digital logic to an Arduino, wiring a small DC motor to a digital logic gate, and programming a dc motor controller function selection cursor for an LCD. Mini lab testing procedures, Arduino code explanation, and the final project assembly will be provided and discussed in this chapter.

2
Programmable DC Motor Controller with an LCD

A **programmable logic controller** (**PLC**) is an industrial computer that is used to operate various electronic and electromechanical devices that are wired to I/O wiring modules. The PLC receives signals from sensors, transducers, and electromechanical switches that are wired to its input wiring module and processes the electrical data by using a microcontroller. The embedded software that is stored in the microcontroller's memory can control external devices, such as electromechanical relays, motors (the AC and DC types), solenoids, and visual displays that are wired to its output wiring module. The PLC programmer programs the industrial computer by using a special programming language known as **ladder logic**. The PLC ladder logic is a graphical programming language that uses computer instruction symbols for automation and controls to operate robots, industrial machines, and conveyor systems.

The PLC, along with the ladder logic software, is very expensive. However, with its off-the-shelf electronic components, Arduino can be used as an alternate mini industrial controller for Maker type robotics and machine control projects. In this chapter, we will see how Arduino can operate as a mini PLC that is capable of controlling a small electric DC motor with a simple two-step programming procedure. Details regarding how one can interface a transistor DC motor with a discrete digital logic circuit to an Arduino and write the control cursor selection code will be provided as well. This chapter will also provide the build instructions for a programmable motor controller. The LCD will provide the programming directions that are needed to operate an electric motor. The parts that are required to build a programmable motor controller are shown in the next section.

Parts list

The following list comprises the parts that are required to build the programmable motor controller:

- Arduino Uno: one unit
- 1 kilo ohm resistor (brown, black, red, gold): three units
- A 10-ohm resistor (brown, black, black, gold): one unit
- A 10-kilo ohm resistor (brown, black, orange, gold): one unit
- A 100-ohm resistor (brown, black, brown, gold): one unit
- A 0.01 µF capacitor: one unit
- An LCD module: one unit
- A 74LS08 Quad AND Logic Gate Integrated Circuit: one unit
- A 1N4001 general purpose silicon diode: one unit
- A DC electric motor (3 V rated): one unit
- Single-pole double-throw (SPDT) electric switches: two units
- 1.5V batteries: two units
- 3V battery holder: one unit
- A breadboard
- Wires

A programmable motor controller block diagram

The block diagram of the programmable DC motor controller with a **Liquid Crystal Display (LCD)** can be imagined as a remote control box with two slide switches and an LCD, as shown in following diagram:

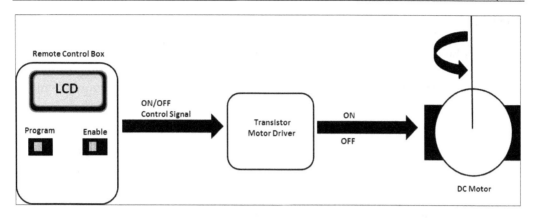

The **Remote Control Box** provides the control signals to operate a DC motor. This box is not able to provide the right amount of electrical current to directly operate the DC motor. Therefore, a transistor motor driver circuit is needed. This circuit has sufficient current gain h_{fe} to operate a small DC motor. A typical hfe value of 100 is sufficient for the operation of a DC motor. The **Enable** slide switch is used to set the remote control box to the `ready` mode. The **Program** switch allows the DC motor to be set to an **ON** or **OFF** operating condition by using a simple selection sequence. The LCD displays the **ON** or **OFF** selection prompts that help you operate the DC motor. The remote control box diagram is shown in the next image.

The idea behind the concept diagram is to illustrate how a simple programmable motor controller can be built by using basic electrical and electronic components. The Arduino is placed inside the remote control box and wired to the Enable/Program switches and the LCD. External wires are attached to the transistor motor driver, DC motor, and Arduino.

As discussed in *Chapter 1, A Sound Effects Machine*, the block diagram of the programmable motor controller is an engineering tool that is used to convey a complete product design by using simple graphics. The block diagram also allows ease in planning the breadboard to prototype and test the programmable motor controller in a maker workshop or a laboratory bench. A final observation regarding the block diagram of the programmable motor controller is that the basic computer convention of inputs is on the left, the processor is located in the middle, and the outputs are placed on the right-hand side of the design layout. As shown, the **SPDT switches** are on the left-hand side, Arduino is located in the middle, and the transistor motor driver with the DC Motor is on the right-hand side of the block diagram.

The LCD is shown towards the right of the block diagram because it is an output device. The LCD allows visual selection between the **ON/OFF** operations of the DC motor by using Program switch. This left-to-right design method allows ease in building the programmable motor controller as well as troubleshooting errors during the testing phase of the project. The block diagram for a programmable motor controller is as follows:

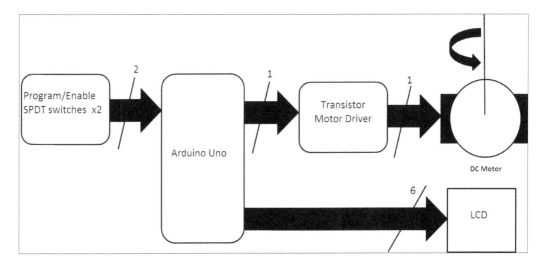

Building the programmable motor controller

The block diagram of the programmable motor controller has more circuits than the block diagram of the sound effects machine that we talked about in *Chapter 1, Sound Effects Machine*. As discussed previously, there are a variety of ways to build the (prototype) electronic devices. For instance, they can be built on a **Printed Circuit Board (PCB)** or an experimenter's/prototype board. The construction base that was used to build this device was a solderless breadboard, which is shown in the next image. The placement of the electronic parts, as shown in the image, are not restricted to the solderless breadboard layout. Rather, it should be used as a guideline.

Another method of placing the parts onto the solderless breadboard is to use the block diagram that was shown earlier. This method of arranging the parts that was illustrated in the block diagram allows ease in testing each subcircuit separately. For example, the **Program/Enable** SPDT switches' subcircuits can be tested by using a DC voltmeter. Placing a DC voltmeter across the Program switch and 1 kilo ohm resistor and toggling switch several times will show a voltage swing between 0 V and +5V. The same testing method can be carried out on the **Enable** switch as well. The transistor motor driver circuit is tested by placing a +5 V signal on the base of the 2N3904 NPN transistor. When you apply +5 V to the transistor's base, the DC motor turns on. The final test for the programmable DC motor controller is to adjust the contrast control (10 kilo ohm) to see whether the individual pixels are visible on the LCD. This electrical testing method, which is used to check the programmable DC motor controller is functioning properly, will minimize the electronic I/O wiring errors. Also, the electrical testing phase ensures that all the I/O circuits of the electronics used in the circuit are working properly, thereby allowing the maker to focus on coding the software. Following is the wiring diagram of programmable DC motor controller with the LCD using a solderless breadboard:

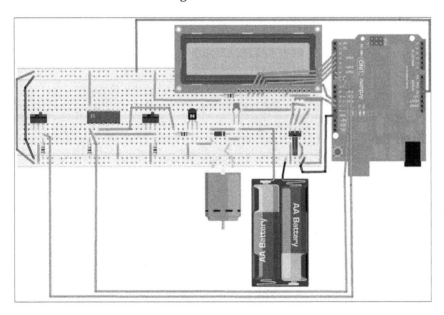

As shown in the wiring diagram, the electrical components that are used to build the programmable DC motor controller with the LCD circuit are placed on the solderless breadboard for ease in wiring the Arduino, LCD, and the DC motor. The transistor shown in the preceding image is a 2N3904 NPN device with a pin-out arrangement consisting of an emitter, a base, and a collector respectively. If the transistor pins are wired incorrectly, the DC motor will not turn on. The LCD module is used as a visual display, which allows operating selection of the DC motor. The program slide switch turns the DC motor ON or OFF. Although most of the 16-pin LCD modules have the same electrical pin-out names, consult the manufacturer's datasheet of the available device in hand. There is also a 10 kilo ohm potentiometer to control the LCD's contrast. On wiring the LCD to the Arduino, supply power to the microcontroller board by using the USB cable that is connected to a desktop PC or a notebook. Adjust the 10 kilo ohm potentiometer until a row of square pixels are visible on the LCD.

The Program slide switch is used to switch between the ON or OFF operating mode of the DC motor, which is shown on the LCD. The 74LS08 Quad AND gate is a 14-pin **Integrated Circuit** (**IC**) that is used to *enable* the DC motor or get the electronic controller *ready* to operate the DC motor. Therefore, the Program slide switch must be in the *ON* position for the electronic controller to operate properly. The 1N4001 diode is used to protect the 2N3904 NPN transistor from peak currents that are stored by the DC motor's winding while turning on the DC motor. When the DC motor is turned off, the 1N4001 diode will direct the peak current to flow through the DC motor's windings, thereby suppressing the transient electrical noise and preventing damage to the transistor. Therefore, it's important to include this electronic component into the design, as shown in the wiring diagram, to prevent electrical damage to the transistor. Besides the wiring diagram, the circuit's schematic diagram will aid in building the programmable motor controller device.

Let's build it!

In order to build the programmable DC motor controller, follow the following steps:

1. Wire the programmable DC motor controller with the LCD circuit on a solderless breadboard, as shown in the previous image as well as the circuit's schematic diagram that is shown in the next image.

2. Upload the software of the programmable motor controller to the Arduino by using the sketch shown next.

3. Close both the Program and Enables switches. The motor will spin.

4. When you open the Enable switch, the motor stops.

The LCD message tells you how one can set the Program switch for an ON and OFF motor control. The Program switch allows you to select between the ON and OFF motor control functions. With the Program switch closed, toggling the Enable switch will turn the motor ON and OFF. Opening the Program switch will prevent the motor from turning on. The next few sections will explain additional details on the I/O interfacing of discrete digital logic circuits and a small DC motor that can be connected to the Arduino.

[A **sketch** is a unit of code that is uploaded to and run on an Arduino board.]

```
/*
 * programmable DC motor controller w/LCD allows the user to select
ON and OFF operations using a slide switch. To
 * enable the selected operation another slide switch is used to
initiate the selected choice.
 * Program Switch wired to pin 6.
 * Output select wired to pin 7.
 * LCD used to display programming choices (ON or OFF).
 * created 24 Dec 2012
 * by Don Wilcher
*/
// include the library code:
#include <LiquidCrystal.h>

// initialize the library with the numbers of the interface pins
LiquidCrystal lcd(12, 11, 5, 4, 3, 2);

// constants won't change. They're used here to
// set pin numbers:
const int ProgramPin = 6;      // pin number for PROGRAM input control
signal
const int OUTPin = 7;        // pin number for OUTPUT control signal

// variable will change:
```

```
int ProgramStatus = 0; // variable for reading Program input
status


void setup() {
  // initialize the following pin as an output:
  pinMode(OUTPin, OUTPUT);

  // initialize the following pin as an input:
  pinMode(ProgramPin, INPUT);

  // set up the LCD's number of rows and columns:
  lcd.begin(16, 2);

  // set cursor for messages andprint Program select messages on
the LCD.
  lcd.setCursor(0,0);
  lcd.print( "1. ON");
  lcd.setCursor(0, 1);
  lcd.print ( "2. OFF");

}

void loop(){
  // read the status of the Program Switch value:
  ProgramStatus = digitalRead(ProgramPin);

  // check if Program select choice is 1.ON.
  if(ProgramStatus == HIGH) {
    digitalWrite(OUTPin, HIGH);

  }
  else{
     digitalWrite(OUTPin,LOW);
  }
}
```

The schematic diagram of the circuit that is used to build the programmable DC motor controller and upload the sketch to the Arduino is shown in the following image:

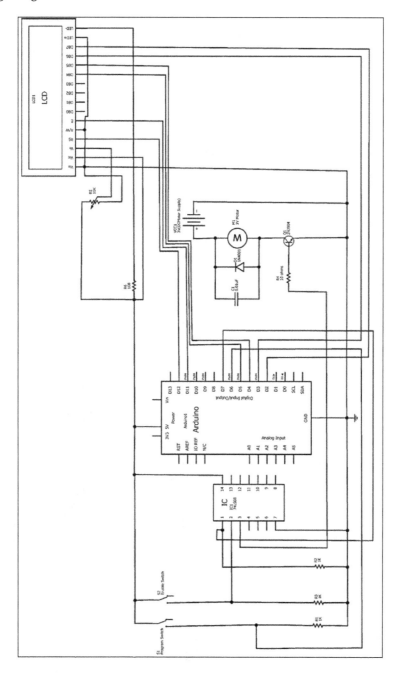

Interfacing a discrete digital logic circuit with Arduino

The Enable switch, along with the Arduino, is wired to a discrete digital **Integrated Circuit (IC)** that is used to turn on the transistor motor driver. The discrete digital IC used to turn on the transistor motor driver is a 74LS08 Quad AND gate. The AND gate provides a high output signal when both the inputs are equal to +5 V. The Arduino provides a high input signal to the 74LS08 AND gate IC based on the following line of code:

```
digitalWrite(OUTPin, HIGH);
```

The OUTPin constant name is declared in the Arduino sketch by using the following declaration statement:

```
const int OUTPin = 7;      // pin number for OUTPUT control signal
```

The Enable switch is also used to provide a +5V input signal to the 74LS08 AND gate IC. The Enable switch circuit schematic diagram is as follows:

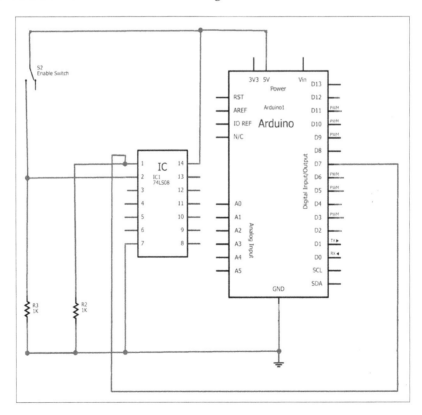

Both the inputs must have the value of logic 1 (+5 V) to make an AND logic gate produce a binary 1 output. In the following section, the truth table of an AND logic gate is given. The table shows all the input combinations along with the resultant outputs. Also, along with the truth table, the symbol for an AND logic gate is provided. A truth table is a graphical analysis tool that is used to test digital logic gates and circuits. By setting the inputs of a digital logic gate to binary 1 (5V) or binary 0 (0 V), the truth table will show the binary output values of 1 or 0 of the logic gate. The truth table of an AND logic gate is given as follows:

AND LOGIC GATE SYMBOL

A

B

C

INPUTS		OUTPUT
A	B	C
0	0	0
0	1	0
1	0	0
1	1	1

Another tool that is used to demonstrate the operation of the digital logic gates is the Boolean Logic expression. The Boolean Logic expression for an AND logic gate is as follows:

$$C = AB \text{ (BOOLEAN LOGIC EXPRESSION)}$$

A Boolean Logic expression is an algebraic equation that defines the operation of a logic gate. As shown for the AND gate, the Boolean Logic expression circuit's output, which is denoted by **C**, is only equal to the product of **A** and **B** inputs. Another way of observing the operation of the AND gate, based on its Boolean Logic Expression, is by setting the value of the circuit's inputs to 1. Its output has the same binary bit value. The truth table graphically shows the results of the Boolean Logic expression of the AND gate.

A common application of the AND logic gate is the *Enable* circuit. The output of the Enable circuit will only be turned on when both the inputs are on. When the Enable circuit is wired correctly on the solderless breadboard and is working properly, the transistor driver circuit will turn on the DC motor that is wired to it. The operation of the programmable DC motor controller's Enable circuit is shown in the following truth table:

INPUTS		OUTPUT
Enable Switch	D7	DC Motor
Open	0V	OFF
Open	+5V	OFF
Closed	0V	OFF
Closed	+5V	ON

The basic computer circuit that makes the decision to operate the DC motor is the AND logic gate. The previous schematic diagram of the Enable Switch circuit shows the electrical wiring to the specific pins of the 74LS08 IC, but internally, the AND logic gate is the main circuit component for the programmable DC motor controller's Enable function. Following is the diagram of 74LS08 AND Logic Gate IC:

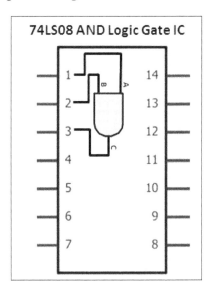

To test the Enable circuit function of the programmable DC motor controller, the Program switch is required. The schematic diagram of the circuit that is required to wire the Program Switch to the Arduino is shown in the following diagram. The Program and Enable switch circuits are identical to each other because two 5 V input signals are required for the AND logic gate to work properly. The Arduino sketch that was used to test the Enable function of the programmable DC motor is shown in the following diagram:

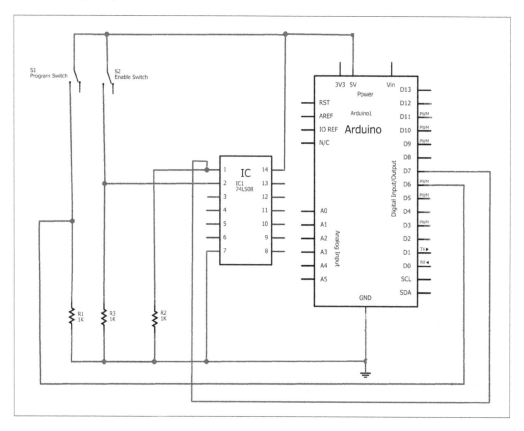

The program for the discrete digital logic circuit with an Arduino is as follows:

```
// constants won't change. They're used here to
// set pin numbers:
const int ProgramPin = 6;   // pin number for PROGRAM input
control signal
const int OUTPin = 7;       // pin number for OUTPUT control signal

// variable will change:
int ProgramStatus = 0;      // variable for reading Program input
status

void setup() {
  // initialize the following pin as an output:
  pinMode(OUTPin, OUTPUT);

  // initialize the following pin as an input:
  pinMode(ProgramPin, INPUT);
}

void loop(){

  // read the status of the Program Switch value:
  ProgramStatus = digitalRead(ProgramPin);

  // check if Program switch is ON.
  if(ProgramStatus == HIGH) {
    digitalWrite(OUTPin, HIGH);

  }
  else{
     digitalWrite(OUTPin,LOW);

  }
}
```

Connect a DC voltmeter's positive test lead to the D7 pin of the Arduino. Upload the preceding sketch to the Arduino and close the Program and Enable switches. The DC voltmeter should approximately read +5 V. Opening the Enable switch will display 0 V on the DC voltmeter. The other input conditions of the Enable circuit can be tested by using the truth table of the AND Gate that was shown earlier. Although the DC motor is not wired directly to the Arduino, by using the circuit schematic diagram shown previously, the truth table will ensure that the programmed Enable function is working properly.

Next, connect the DC voltmeter to the pin 3 of the 74LS08 IC and repeat the truth table test again. The pin 3 of the 74LS08 IC will only be ON when both the Program and Enable switches are closed. If the AND logic gate IC pin generates wrong data on the DC voltmeter when compared to the truth table, recheck the wiring of the circuit carefully and properly correct the mistakes in the electrical connections. When the corrections are made, repeat the truth table test for proper operation of the Enable circuit.

Interfacing a small DC motor with a digital logic gate

The 74LS08 AND Logic Gate IC provides an electrical interface between reading the Enable switch trigger and the Arduino's digital output pin, pin D7. With both the input pins (1 and 2) of the 74LS08 AND logic gate set to binary 1, the small 14-pin IC's output pin 3 will be High. Although the logic gate IC's output pin has a +5 V source present, it will not be able to turn a small DC motor. The 74LS08 logic gate's sourcing current is not able to directly operate a small DC motor.

To solve this problem, a transistor is used to operate a small DC motor. The transistor has sufficient current gain hfe to operate the DC motor. The DC motor will be turned on when the transistor is biased properly. **Biasing** is a technique pertaining to the transistor circuit, where providing an input voltage that is greater than the base-emitter junction voltage (VBE) turns on the semiconductor device. A typical value for VBE is 700 mV. Once the transistor is biased properly, any electrical device that is wired between the collector and +VCC (collector supply voltage) will be turned on. An electrical current will flow from +VCC through the DC motor's windings and the collector-emitter is grounded.

The circuit that is used to operate a small DC motor is called a **Transistor motor driver**, and is shown in the following diagram:

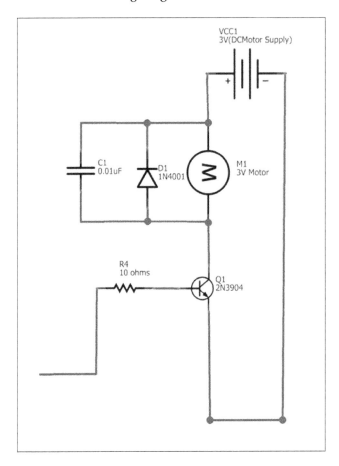

The Arduino code that is responsible for the operation of the transistor motor driver circuit is as follows:

```
void loop(){

    // read the status of the Program Switch value:
    ProgramStatus = digitalRead(ProgramPin);

    // check if Program switch is ON.
    if(ProgramStatus == HIGH) {
        digitalWrite(OUTPin, HIGH);

    }
}
```

```
else{
    digitalWrite(OUTPin,LOW);

    }
}
```

Although the transistor motor driver circuit was not directly wired to the Arduino, the output pin of the microcontroller prototyping platform indirectly controls the electromechanical part by using the 74LS08 AND logic gate IC. A tip to keep in mind when using the transistors is to ensure that the semiconductor device can handle the current requirements of the DC motor that is wired to it. If the DC motor requires more than 500 mA of current, consider using a power **Metal Oxide Semiconductor Field Effect Transistor (MOSFET)** instead.

A power MOSFET device such as IRF521 (N-Channel) and 520 (N-Channel) can handle up to 1 A of current quite easily, and generates very little heat. The low heat dissipation of the **power MOSFET (PMOSFET)** makes it more ideal for the operation of high-current motors than a general-purpose transistor. A simple PMOSFET DC motor driver circuit can easily be built with a handful of components and tested on a solderless breadboard, as shown in the following image. The circuit schematic for the solderless breadboard diagram is shown after the breadboard image as well. **Sliding the Single Pole-Double Throw (SPDT)**switch in one position biases the PMOSFET and turns on the DC motor. Sliding the switch in the opposite direction turns off the PMOSFET and the DC motor.

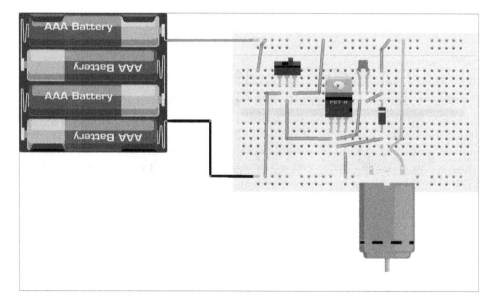

> Once this circuit has been tested on the solderless breadboard, replace the 2N3904 transistor in the programmable DC Motor controller project with the power-efficient PMOSFET component mentioned earlier.

As an additional reference, the schematic diagram of the transistor relay driver circuit is as follows:

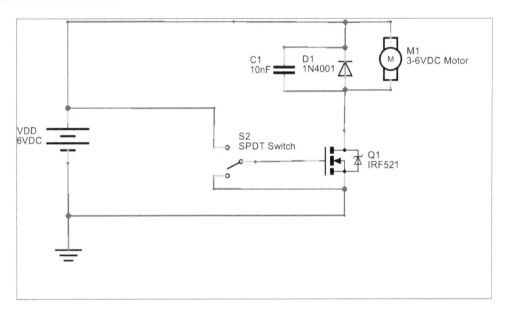

A sketch of the LCD selection cursor

The LCD provides a simple user interface for the operation of a DC motor that is wired to the Arduino-based programmable DC motor controller. The LCD provides the two basic motor operations of *ON* and *OFF*. Although the LCD shows the two DC motor operation options, the display doesn't provide any visual indication of selection when using the Program switch. An enhancement feature of the LCD is that it shows which DC motor operation has been selected by adding a selection symbol. The LCD selection feature provides a visual indicator of the DC motor operation that was selected by the Program switch. This selection feature can be easily implemented for the programmable DC motor controller LCD by adding a > symbol to the Arduino sketch.

After uploading the original sketch from the *Let's build it* section of this chapter, the LCD will display two DC motor operation options, as shown in the following image:

The enhancement concept sketch of the new LCD selection feature is as follows:

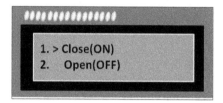

The selection symbol points to the DC motor operation that is based on the Program switch position. (For reference, see the schematic diagram of the programmable DC motor controller circuit.)

The partially programmable DC motor controller program sketch that comes without an LCD selection feature

Comparing the original LCD DC motor operation selection with the new sketch, the differences with regard to the programming features are as follows:

```
void loop(){

    // read the status of the Program Switch value:
    ProgramStatus = digitalRead(ProgramPin);

    // check if Program switch is ON.
    if(ProgramStatus == HIGH) {
      digitalWrite(OUTPin, HIGH);

    }
    else{
```

```
        digitalWrite(OUTPin,LOW);


    }
}
```

The partially programmable DC motor controller program sketch with an LCD selection feature

This code feature will provide a selection cursor on the LCD to choose the programmable DC motor controller operation mode:

```
   // set cursor for messages and print Program select messages on the
LCD.
     lcd.setCursor(0,0);
     lcd.print( ">1.Closed(ON)");
     lcd.setCursor(0, 1);
     lcd.print ( ">2.Open(OFF)");


 void loop(){

   // read the status of the Program Switch value:
   ProgramStatus = digitalRead(ProgramPin);

   // check if Program select choice is 1.ON.
   if(ProgramStatus == HIGH) {
     digitalWrite(OUTPin, HIGH);
       lcd.setCursor(0,0);
       lcd.print( ">1.Closed(ON)");
       lcd.setCursor(0,1);
       lcd.print ( " 2.Open(OFF) ");

   }
   else{
       digitalWrite(OUTPin,LOW);
        lcd.setCursor(0,1);
        lcd.print ( ">2.Open(OFF)");
        lcd.setCursor(0,0);
        lcd.print( " 1.Closed(ON) ");

   }
}
```

The most obvious difference between the two partial Arduino sketches is that the LCD selection feature has several lines of code as compared to the original one. As the slide position of the Program switch changes, the LCD's selection symbol instantly moves to the correct operating mode. Although the DC motor can be observed directly, the LCD confirms the operating mode of the electromechanical device. The complete LCD selection sketch is shown in the following section. As a design-related challenge, try displaying an actual arrow for the DC motor operating mode on the LCD. As illustrated in the sketch, an arrow can be built by using the keyboard symbols or the **American Standard Code for Information Interchange (ASCII)** code.

```
/*
 *   programmable DC motor controller w/LCD allows the user to select
ON and OFF operations using a slide switch. To
 *   enable the selected operation another slide switch is used to
initiate the selected choice.
 *   Program Switch wired to pin 6.
 *   Output select wired to pin 7.
 *   LCD used to display programming choices (ON or OFF) with selection
arrow.
 *   created 28 Dec 2012
 *   by Don Wilcher
 */
// include the library code:
#include <LiquidCrystal.h>

// initialize the library with the numbers of the interface pins
LiquidCrystal lcd(12, 11, 5, 4, 3, 2);

// constants won't change. They're used here to
// set pin numbers:
const int ProgramPin = 6;     // pin number for PROGRAM input control
signal
const int OUTPin = 7;         // pin number for OUTPUT control signal


// variable will change:
int ProgramStatus = 0;        // variable for reading Program input
status


void setup() {
  // initialize the following pin as an output:
```

```
    pinMode(OUTPin, OUTPUT);

    // initialize the following pin as an input:
    pinMode(ProgramPin, INPUT);

    // set up the LCD's number of rows and columns:
    lcd.begin(16, 2);

    // set cursor for messages andprint Program select messages on the
LCD.
    lcd.setCursor(0,0);
    lcd.print( ">1.Closed(ON)");
    lcd.setCursor(0, 1);
    lcd.print ( ">2.Open(OFF)");

}

void loop(){

    // read the status of the Program Switch value:
    ProgramStatus = digitalRead(ProgramPin);

    // check if Program select choice is 1.ON.
    if(ProgramStatus == HIGH) {
      digitalWrite(OUTPin, HIGH);
        lcd.setCursor(0,0);
        lcd.print( ">1.Closed(ON)");
        lcd.setCursor(0,1);
        lcd.print ( " 2.Open(OFF) ");

    }
    else{
        digitalWrite(OUTPin,LOW);
         lcd.setCursor(0,1);
         lcd.print ( ">2.Open(OFF)");
         lcd.setCursor(0,0);
         lcd.print( " 1.Closed(ON) ");

    }
}
```

Congratulations on building your programmable motor controller device!

Summary

In this chapter, a programmable motor controller was built by using an Arduino, AND gate, and transistor motor driver. The fundamentals of digital electronics, which include the concepts of Boolean logic expressions and truth tables were explained in the chapter. The AND gate is not able to control a small DC motor because of the high amount of current that is needed to operate it properly. PMOSFET (IRF521) is able to operate a small DC motor because of its high current sourcing capability. The circuit that is used to wire a transistor to a small DC motor is called a transistor DC motor driver. The DC motor can be turned on or off by using the LCD cursor selection feature of the programmable DC motor controller.

In *Chapter 3*, *Talking Logic Probe*, we will have a discussion on how to make an electronic tester that gives an output of the binary state (high or low) of a digital logic gate circuit by using a programmable text-to-speech module. EMIC 2, a programmable text-to-speech module will be introduced in this chapter. In addition to the EMIC 2 programmable text-to-speech module, the following hands-on activities will be presented:

- Installing EMIC 2 library to your Arduino sketch (code)
- How to use the EMIC 2 Arduino commands
- Wiring the EMIC 2 programmable text-to-speech module to an Arduino
- Adding a small transistor amplifier circuit for speech output
- How to use the talking logic probe to test a digital logic gate circuit

Mini lab testing procedures, an explanation of the Arduino code, and the final project assembly will be provided and discussed in the next chapter.

3

A Talking Logic Probe

Digital circuits are electronic devices that show either high or low operating states. Other ways of describing digital circuit operating states include true or false and binary 1 or 0. A **Truth Table (TT)**, as discussed in the previous chapter, is a graphical tool used to test the electrical operation of digital circuits by showing input data and observing their output states. The TT aids the makers building digital circuits by allowing them to inject test data into the device's input pins and observing the output states using a logic probe.

A logic probe is an electrical tester that displays input and output signals of a digital circuit using visual indicators. LEDs are typical visual indicators used to show the binary data of a digital circuit. Typical colors used to represent high and low digital data are red and green, respectively. Some variations of logic probe visual indicator may consist of a seven segment display capable of displaying the letters H or L for high or low binary data, respectively. A typical logic probe to test digital circuits is shown here:

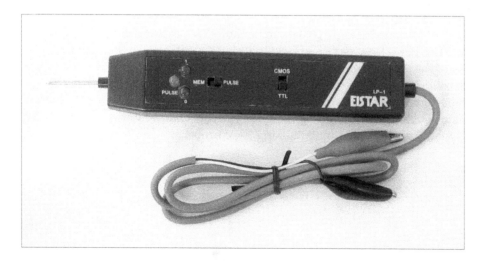

[Binary switches provide input data for digital circuits.]

In this chapter, we will learn how to build a talking logic probe for testing digital circuits and devices. Also, to make the logic probe unique from the typical off-the-shelf testers, voice synthesis will be used to speak the operating conditions of digital circuits. An EMIC 2 **text-to-speech** (**TTS**) module will be used to generate the voice synthesis for the logic probe. Specific EMIC 2 TTS module operation modes will be explored in this chapter using different Arduino sketches. This chapter will also provide instructions for building the logic probe.

Parts list

The following is the list of parts required for building a talking logic probe:

- Arduino Uno (one unit)
- (1) EMIC 2 text-to-speech module (one unit)
- 8 ohm speaker (one unit)
- 1 kilo ohm resistor (black, brown, red, and gold) (one unit)
- Momentary pushbutton electrical switch (one unit)
- Breadboard
- Wires

A talking logic probe block diagram

The concept of a talking logic probe can be thought of as a typical testing device with a speaker. The LEDs for binary high and low status can be removed from the logic probe because of the voice synthesis feature of the digital tester. To test different digital IC families with a logic probe is very important in electronics because today's products use them extensively. Therefore, a slide switch for selecting between **Transistor-Transistor Logic** (**TTL**) and **Complementary Metal-Oxide-Semiconductor** (**CMOS**) type digital ICs is provided in the talking logic probe.

CMOS devices use FETs (Field Effect Transistors) instead of Bipolar Junction transistors (BJTs) used by TTL ICs. The CMOS technology is being used in all the semiconductor devices because of lower power dissipation as compared to the TTL technologies. Also, some CMOS ICs can operate on as low as 1.8 V DC to 2.0V DC whereas TTL devices supply a voltage range between 3.75V DC to 5.75V DC.

A concept diagram of a talking logic probe is shown here:

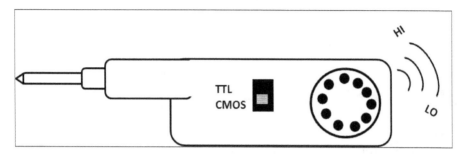

The talking logic probe block diagram is an engineering development tool used to convey a complete product design using simple graphics. The block diagram also makes it easier to plan the breadboard for prototyping and testing of the talking logic probe in a maker's workshop or laboratory bench. A final observation of the talking logic probe block diagram is that the basic computer convention of inputs is on the left-hand side, the processor is located in the middle, and the outputs are placed on the right-hand side of the design layout. As shown, the **Digital Circuit under Test (DCuT)** is on the left-hand side, the Arduino is located in the middle, and the EMIC 2 TTS module with the 8 ohm speaker is shown on the right-hand side of the block diagram. The DCuT will provide binary high or low signals, based on the digital circuits inputs being triggered.

This left to right design method makes it easier to build the talking logic probe and troubleshoot the errors during the testing phase of the project.

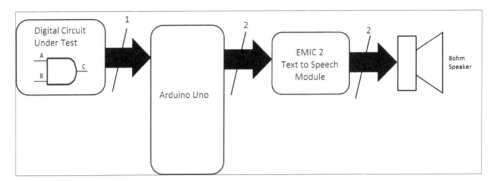

A talking logic probe – testing the EMIC 2 TTS module

The talking logic probe block diagram is simple in design as compared to the block diagram we used in *Chapter 2, Programmable DC Motor Controller with LCD*. As discussed in *Chapter 1, A Sound Effects Machine*, there are a variety of ways to build (prototype) electronic devices, such as on a **Printed Circuit Board** (**PCB**) or an experimenter/prototype board. The prototyping tool used to build and test the device's EMIC 2 TTS module is a solderless breadboard shown next. The placement of electronic parts, as shown in the following diagram, is not restricted to the solderless breadboard layout, but is used as a guideline:

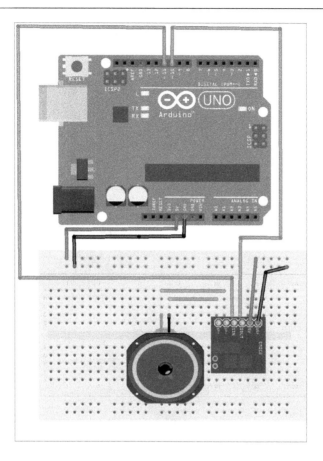

Another method of parts placement onto the solderless breadboard is to use the block diagram shown previously. This method of parts arrangement, illustrated in the block diagram, allows ease in testing each subcircuit separately. For example, the EMIC 2 TTS module can be tested by using a sample code given next. With the sketch uploaded to the Arduino, the musical song *Daisy Bell* can be heard through the speaker. After the song has played once, pressing the reset button repeats the musical lyrics:

```
/*
   Emic 2 Text-to-Speech Module: Basic Demonstration

   Author: Joe Grand [www.grandideastudio.com]
   Contact: support@parallax.com
```

```
Program Description:

This program provides a simple demonstration of the Emic 2 Text-
to-Speech
Module. Please refer to the product manual for full details of
system
functionality and capabilities.

Revisions:

1.0 (February 13, 2012): Initial release
1.1 (April 29, 2014): Changed rxPin/txPin to use pins 10/11,
respectively, for widest support across the Arduino family
(http://arduino.cc/en/Reference/SoftwareSerial)

*/

// include the SoftwareSerial library so we can use it to talk to
the Emic 2 module
#include <SoftwareSerial.h>

#define rxPin    10  // Serial input (connects to Emic 2's SOUT
pin)
#define txPin    11  // Serial output (connects to Emic 2's SIN
pin)
#define ledPin   13  // Most Arduino boards have an on-board LED on
this pin

// set up a new serial port
SoftwareSerial emicSerial = SoftwareSerial(rxPin, txPin);

void setup()  // Set up code called once on start-up
{
  // define pin modes
  pinMode(ledPin, OUTPUT);
  pinMode(rxPin, INPUT);
  pinMode(txPin, OUTPUT);

  // set the data rate for the SoftwareSerial port
  emicSerial.begin(9600);

  digitalWrite(ledPin, LOW);  // turn LED off
```

```
/*
   When the Emic 2 powers on, it takes about 3 seconds for it to
   successfully
   initialize. It then sends a ":" character to indicate it's
   ready to accept
   commands. If the Emic 2 is already initialized, a CR will also
   cause it
   to send a ":"
*/
   emicSerial.print('\n');              // Send a CR in case the
   system is already up
   while (emicSerial.read() != ':');    // When the Emic 2 has
   initialized and is ready, it will send a single ':' character,
   so wait here until we receive it
   delay(10);                           // Short delay
   emicSerial.flush();                  // Flush the receive buffer
}

void loop()  // Main code, to run repeatedly
{
   // Speak some text
   emicSerial.print('S');
   emicSerial.print("Hello. My name is the Emic 2 Text-to-Speech
   module. I would like to sing you a song.");  // Send the desired
   string to convert to speech
   emicSerial.print('\n');
   digitalWrite(ledPin, HIGH);          // Turn on LED while Emic is
   outputting audio
   while (emicSerial.read() != ':');    // Wait here until the Emic
   2 responds with a ":" indicating it's ready to accept the next
   command
   digitalWrite(ledPin, LOW);

   delay(500);     // 1/2 second delay

   // Sing a song
   emicSerial.print("D1\n");
   digitalWrite(ledPin, HIGH);          // Turn on LED while Emic is
   outputting audio
   while (emicSerial.read() != ':');    // Wait here until the Emic
   2 responds with a ":" indicating it's ready to accept the next
   command
   digitalWrite(ledPin, LOW);
```

```
while(1)        // Demonstration complete!
{
  delay(500);
  digitalWrite(ledPin, HIGH);
  delay(500);
  digitalWrite(ledPin, LOW);
}
}
```

In addition to the test sketch, the circuit schematic diagram for the solderless breadboard wiring diagram is shown for reference:

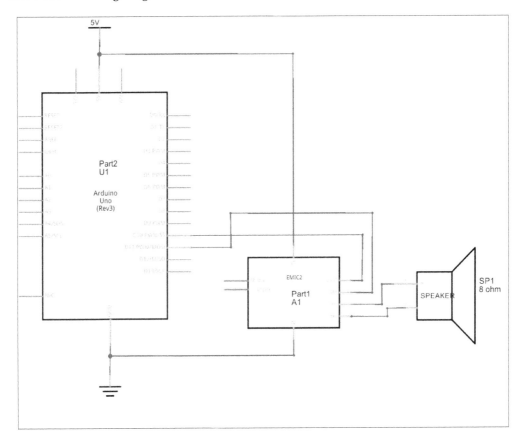

EMIC 2 TTS module basics

The heart of the talking logic probe is the voice synthesis engine for speaking the binary high and low signals produced by the digital circuits. The EMIC 2 TTS module, as shown next, is a versatile, multi language voice synthesizer capable of taking digital text and converting it into audible speech. The EMIC 2 TTS module uses the DECtalk text for text-to-speech synthesizer engine. As shown in the circuit schematic and breadboard wiring diagrams, the EMIC 2 TTS module is quite easy to wire to the Arduino. This ease of adding an audible voice to electrical testers can provide an eyes-free approach to troubleshooting electronic circuits. Instead of looking at a visual display for measurement data, the EMIC 2 TTS can provide speech output for the electrical measuring device. This section of the chapter will discuss the basic features and electrical connections of the EMIC 2 TTS module.

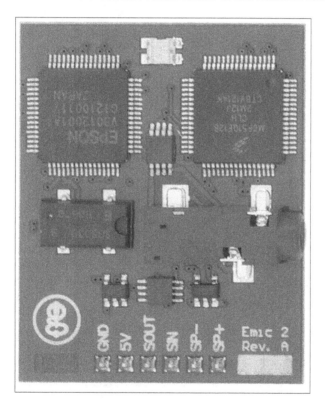

 The EMIC 2 TTS module is manufactured by Parallax Inc and designed by Joe Grand of Grand Design Studio (www.grandideastudio.com).

EMIC 2 TTS module's key features

The EMIC 2 TTS module provides a variety of features that make it appealing to embedded applications, such as robotics, healthcare, automotive, and industrial products. Some key features of the EMIC 2 TTS module include high quality speech synthesis, pre-defined voice styles, an on-board audio amplifier, and a single row header for electrical wiring. The EMIC 2 only requires a +5V DC, 30 mA DC power supply. The operating temperature range of the EMIC 2 TTS module is -20° C to 70° C (-4° F to 150° F), making it suitable for typical environments. Dimensions of the EMIC 2 TTS module are 1.25" w x 1.5" l x 0.37" h, which allows it to fit easily inside a small plastic hobby box. The following are the picture views of the EMIC 2 TTS module:

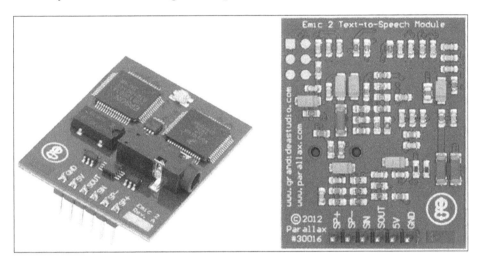

Another thing to note about the EMIC 2 is the small audio jack soldered onto the PCB. This standard audio jack (size 1/8" or 3.5 mm) allows the device to be connected to most audio Hi-Fi systems or headphones, if desired. An example design application for this audio jack is the creation of a portable foreign language training tool. Connection to headphones instead could provide individual training to the foreign language learner without disturbing people in the room. The EMIC 2 TTS module has two speech synthesis languages available for English and Spanish. Another training aid application suitable for the audio jack feature is a physical digital logic simulator with audible I/O feedback. The audio jack could also be used in this project instead of the speaker.

[If both 3.5 mm jack and external 8 ohm speaker are used at the same time, audio quality might be affected.]

Electrical connections

As seen previously, there are only six electrical connections or pins required for wiring the EMIC 2 TTS module to an Arduino. The electrical connections consist of GND, 5 V, SOUT, SIN, SP-, and SP+. The following descriptions explain the operation of these six electrical connections:

- **Pin 1:GND**: This is a device ground pin. Connect the Arduino's supply voltage ground (GND) to this pin.

- **Pin 2: 5 V**: This is a device power pin. Connect the Arduino's positive voltage (5 V) to this pin.

- **Pin 3: SOUT**: This pin is a device serial output to the Arduino. It is a 5 V TTL level interface, 9,600 bps*, 8 data bits, no parity, 1 stop bit, non-inverted digital signal.

- **Pin 4: SIN**: This pin is a device serial input from the Arduino. It is a 3.3 V to 5 V TTL level interface, 9,600 bps, 8 data bits, no parity, 1 stop bit, non-inverted digital signal.

- **Pin 5: SP-**: Device differential audio amplifier output, bridge-tied load configuration, negative side. Connect to 8 ohm speaker directly.

- **Pin 6: SP+**: Device differential audio amplifier output, bridge-tied load configuration, positive side. Connect to 8 ohm speaker directly.

 For a typical serial asynchronous communications rate, the value is 9,600 **bps** (**bits per second**).

The following circuit schematic diagram illustrates this wiring connection scheme:

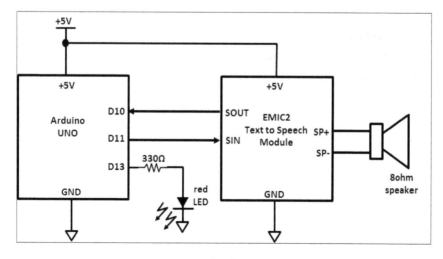

Let's build it!

Now is the time for building our own talking logic probe. Follow these steps:

1. Wire the talking logic probe circuit on a solderless breadboard, as shown next. For reference, the circuit schematic diagram has been provided here:

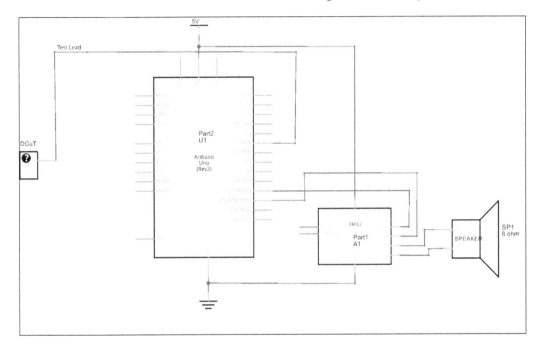

2. Upload the talking logic probe code to the Arduino using the sketch given after the schematic diagram.

3. The phrase *Signal is Low* will be heard through the talking logic probe's speaker.

4. Touch the logic probe's green test lead onto the EMIC 2 TTS module's +5 V pin.

5. The phrase *Signal is High* will be heard through the talking logic probe's speaker.

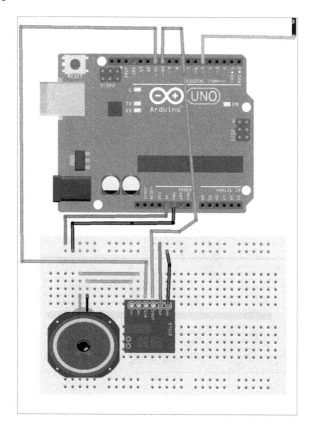

 The mystery part represents the DCuT.

The code required to make the talking logic probe operational is:

```
/*

    Talking Logic Probe

    Author: Don Wilcher 2014 14 11

    Program Description:

    This program allows tracing digital circuit signals using the
    Emic 2 Text-to-Speech
```

```
      Module. Detecting a high or low signal will allow the Emic 2 TTS
      module to speak one of the voltage levels present.


  */


  // include the SoftwareSerial library so we can use it to talk to
  the Emic 2 module
  #include <SoftwareSerial.h>

  #define rxPin   10  // Serial input (connects to Emic 2's SOUT
  pin)
  #define txPin   11  // Serial output (connects to Emic 2's SIN
  pin)
  #define probein 5  // Digital probe input
  int probe=0; // probein variable status

  // set up a new serial port
  SoftwareSerial emicSerial = SoftwareSerial(rxPin, txPin);

  void setup()  // Set up code called once on start-up
  {
    // define pin modes
    pinMode(rxPin, INPUT);
    pinMode(txPin, OUTPUT);
    pinMode(probein, INPUT);

    // set the data rate for the SoftwareSerial port
    emicSerial.begin(9600);

    emicSerial.print('\n');              // Send a CR in case the
    system is already up
    while (emicSerial.read() != ':');   // When the Emic 2 has
    initialized and is ready, it will send a single ':' character,
    so wait here until we receive it
    delay(10);                          // Short delay
    emicSerial.flush();                 // Flush the receive buffer
  }

  void loop()  // Main code, to run repeatedly
  {
    // Speak some text
    emicSerial.print('S');
    probe=digitalRead(probein);// read probein status
    if(probe == HIGH){              // Emic 2 will speak HIGH message if
      status value is true
      emicSerial.print("SIGNAL IS HIGH");  // Send the desired
      string to convert to speech
      emicSerial.print('\n');
```

```
    while (emicSerial.read() != ':');   // Wait here until the
    Emic 2 responds with a ":" indicating it's ready to accept the
    next command
  }
  else{
    emicSerial.print("SIGNAL IS LOW");  // Emic 2 will speak LOW
    message if status value is false
    emicSerial.print('\n');
    while (emicSerial.read() != ':');   // Wait here until the
    Emic 2 responds with a ":" indicating it's ready to accept the
    next command

    delay(1000);    //  1 second delay
  }
}
```

As the talking logic probe toggles between two phrases, based on the detected input signals, occasionally, a text phrase can be out of sync. For example, the SIGNAL IS HIGH phrase may be heard through the speaker with no +5 V DC input signal detected by the Arduino. This unstable condition occurs when a digital or microcontroller's assigned pin isn't properly terminated. Electronic circuits such as a microcontroller's external crystal oscillator can produce radiated switching signals capable of disrupting its operating embedded code. The unstable speech heard through the speaker is, again, produced by the crystal oscillator's radiated switching signal. Removing this electrical noise consists of providing a signal path to ground using a pull-down resistor. Adding a 1 kilo ohm resistor between the Arduino's *D5* pin to ground will provide a stable speech output for the talking logic probe. The solderless breadboard wiring diagram and circuit schematic diagram show a 1 kilo ohm resistor added to the Arduino's *D5* pin.

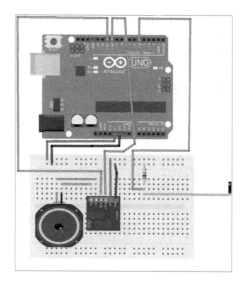

For reference, the following is the talking logic probe circuit schematic diagram:

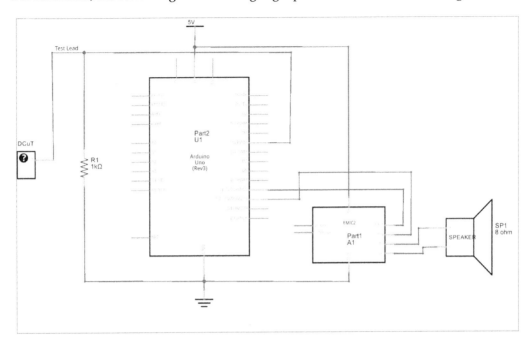

 If a 1 kilo ohm resistor is not available, a 10 kilo ohm resistor may be used as a replacement pull-down resistor.

Another feature that can be added to the talking logic probe is a momentary pushbutton electrical switch for activating the detection function. With the test lead attached to the digital circuit's specific I/O pin, a press of the momentary pushbutton switch will provide a quick go-no-go check of the device's electrical operation. To include this manual signal detection feature requires the momentary pushbutton switch to be wired between the logic probe's test lead and the Arduino's *D5* pin.

This wiring connection is also known as a series circuit. The following is the solderless breadboard wiring diagram for adding the momentary pushbutton electrical switch:

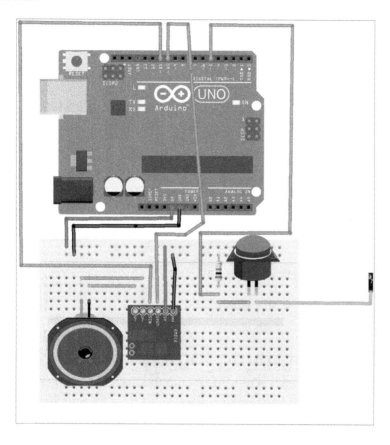

The following is the talking logic probe with a switch detection feature circuit schematic diagram:

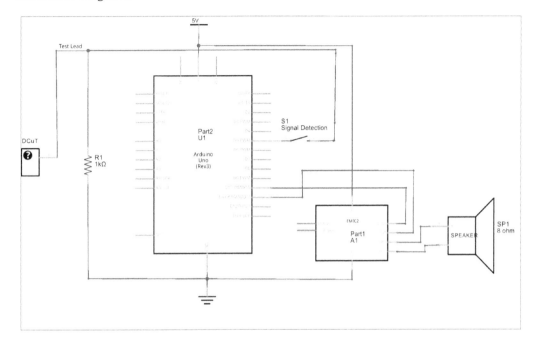

How does the talking logic probe code work

The talking logic probe sketch is modified from the EMIC 2 sample code given earlier in this chapter. The library necessary for the EMIC 2 TTS module to work correctly is:

```
#include <SoftwareSerial.h>
```

Removing this library (header file) from the sketch will cause compilation errors throughout the code, and not allow the Arduino to correctly operate the EMIC 2 TTS module. The pins needed to send digital data between the Arduino and the EMIC 2 TTS module are:

```
#define rxPin   10  // Serial input (connects to Emic 2's SOUT
pin)
#define txPin   11  // Serial output (connects to Emic 2's SIN
pin)
```

The traditional serial communication pins on the Arduino are *D0* and *D1*. These digital pins may be used in the talking logic probe sketch, but must be removed while uploading the sketch. The reason is because the physical link between the Arduino and the EMIC 2 TTS module while uploading a sketch requires using pins *D0* and *D1*. There is an electrical conflict in establishing communications between the desktop PC or notebook computer and the EMIC 2 using the true *tx:D0* and *rx:D1* pins of the Arduino. To illustrate this electrical communication conflict, change the rx and tx pins to *D0* and *D1* on the Arduino. Upload the modified sketch to the Arduino and notice the following error displayed on the IDE's message window:

```
Uploading...

Global variables use 153 bytes (7%) of dynamic memory, leaving
1,895 bytes for local variables. Maximum is 2,048 bytes.
avrdude: stk500_getsync(): not in sync: resp=0x00

20                                              Arduino Uno on COM5
```

To troubleshoot this problem, remove the two wires connected to the Arduino *tx* and *rx* pins while uploading. The wires can be reconnected to these pins after the sketch has been uploaded to the Arduino.

The next two lines of code to examine are as follows:

```
#define probein 5   // Digital probe input
int probe=0; // probein variable status
```

The talking probe test lead is connected to pin *D5* of the Arduino. The digital signals are received using this pin. The `int probe` = 0 status variable will be used to collect the test lead's digital signals during circuit testing. With the test lead pin defined with its signal status variable, the declared Arduino pins operating modes will be set up using the following lines of code:

```
// define pin modes
pinMode(rxPin, INPUT);
pinMode(txPin, OUTPUT);
pinMode(probein, INPUT);
```

The `void setup()` function is used to establish the I/O modes for the declared Arduino pins. The next few lines of code help establish the communication protocol (rules) between the Arduino and EMIC 2 TTS module:

```
// set the data rate for the SoftwareSerial port
emicSerial.begin(9600);

emicSerial.print('\n');               // Send a CR in case the
system is already up
while (emicSerial.read() != ':');    // When the Emic 2 has
initialized and is ready, it will send a single ':' character,
so wait here until we receive it
delay(10);                            // Short delay
emicSerial.flush();                   // Flush the receive buffer
```

The preceding lines of code are required when using the EMIC 2 TTS module for different voice/speech synthesis projects. For creating the talking logic probe sketch, a few lines of code were taken from the EMIC 2 sample sketch from the Parallax Inc website. Code reuse, as illustrated here, saves time in developing Arduino sketches and allows the maker to focus on building the electronics instead of debugging software.

The main `void main ()` code allows the Arduino to read the digital signal detected and speak the voltage level phrases. The first line of code enables the EMIC 2 for speech:

```
// Speak some text
  emicSerial.print('S');
```

With the text-to-speech module ready to speak, the test lead's status can be read using `probein`, and stored in the variable `int probe`:

```
probe=digitalRead(probein);// read probein status
```

Once the signal level has been stored in the status variable probe, determining which message to speak is based on this conditional statement:

```
if(probe == HIGH){           // Emic 2 will speak HIGH message if
status value is true
```

If the statement is true, the EMIC 2 TTS module will speak the phrase SIGNAL IS HIGH using the following line of code:

```
emicSerial.print("SIGNAL IS HIGH");   // Send the desired
string to convert to speech
```

Once this text is converting into speech by the EMIC 2, a carriage return is needed to speak the next line of text:

```
micSerial.print('\n');
```

The next action required is to set up handshaking (communications acknowledgement), where the EMIC 2 waits for a new text message to be sent by the Arduino for speech conversion. Here's the line of code required for the Arduino to EMIC 2 TTS module handshaking:

```
while (emicSerial.read() != ':');   // Wait here until the
Emic 2 responds with a ":" indicating it's ready to accept the
next command
```

The next message phrase for speech conversion when the signal detected by the test lead is low is SIGNAL IS LOW. The else keyword completes the conditional statement instruction of the Arduino code given here:

```
else{
    emicSerial.print("SIGNAL IS LOW");   // Emic 2 will speak
    LOW message if status value is false
```

The two processes of waiting on the next phrase message for conversion and establishing handshaking, as previously discussed, are as follows:

```
emicSerial.print("SIGNAL IS LOW");   // Emic 2 will speak LOW
message if status value is false
emicSerial.print('\n');
while (emicSerial.read() != ':');   // Wait here until the Emic
2 responds with a ":" indicating it's ready to accept the next
command
```

The delay (1000) instruction allows the two phrases to repeat every one second, as shown in the last line of Arduino code. The following flowchart graphically summarizes the talking logic probe sketch. The structure of this flowchart can serve as a template when developing EMIC 2 applications. The final section of the chapter will end with a discussion on DecTalk speech synthesizer engine. Code examples of how to set various feature/functions of the EMIC 2 TTS module will be provided for further hands-on exploration.

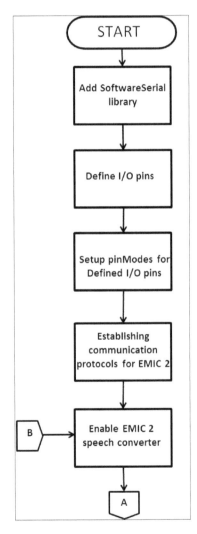

The continuation of the flow chart is shown as follows:

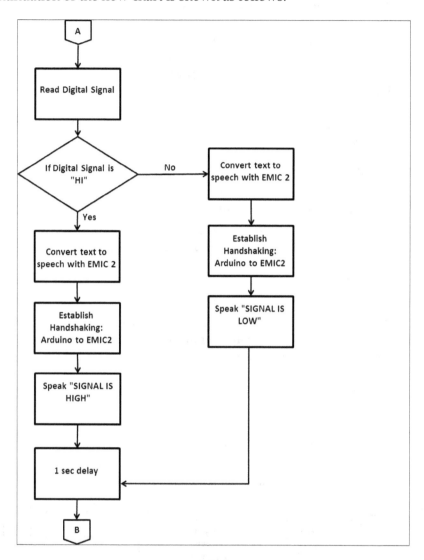

DecTalk speech synthesizer engine

The EMIC 2 TTS module uses a speech synthesizer called **DecTalk**. The DecTalk synthesizer is capable of producing a natural sounding voice based on the smallest memory IC footprint. The synthesizer is able to speak in multilanguages, supporting US English, Castilian, and Latin American Spanish. The 12-page document that Parallax Inc developed provides a wealth of information on maximizing the features of the EMIC 2 TTS module. Following are the examples from the Parallax document on how to set the internal feature/functions of the EMIC 2 using the DecTalk speech synthesizer and Arduino code:

- **Vx**: This sets the audio volume in decibels(dB).

  The audio output volume of the EMIC 2 TTS module can be set in the range of 48 dB (softest) to 18 dB (loudest) using the following Arduino code (the default volume setting is 0):

  ```
  // Set volume to x: x= -48 to 18(max volume)
  emicSerial.print("V5\n");
  while (emicSerial.read() != ':');   // wait for ':'
  character
  ```

  The speech output is sent to a bridged audio power amplifier. A Texas Instruments LM4864 bridged audio power amplifier is used to drive an 8 ohm speaker with an amplification gain (AV) of 10. The LM4864 bridged audio power amplifier circuit is shown here:

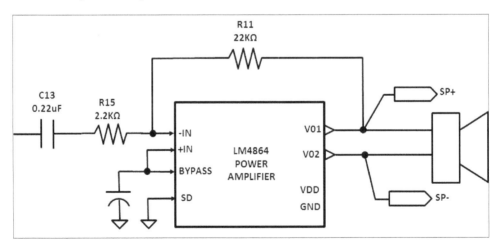

With this audio power amplifier IC and the DecTalk volume command, there is no need to add an amplifier circuit to the EMIC 2 TTS module to obtain a suitable output level.

- **Nx**: This selects an EMIC 2 speaking voice. There are nine voices to choose from as listed here:

    1. Perfect Paul (Paulo)
    2. Huge Harry (Francisco)
    3. Beautiful Betty
    4. Uppity Ursula
    5. Doctor Dennis (Enrique)
    6. Kit the Kid
    7. Frail Frank
    8. Rough Rita
    9. Whispering Wendy (Beatriz)

The EMIC 2 TTS module's voice can be changed using the following Arduino code:

```
// Set voice
  emicSerial.print("N1\n");
  while (emicSerial.read() != ':');   // wait for ':'
  character
```

After uploading the modified talking logic probe sketch with the DecTalk set voice command, Huge Harry (Francisco) will be heard through the speaker.

- **Wx**: This sets Set the EMIC 2 TTS module's speaking rate (words per minute).

Depending on the application, the speaking rate may need to be adjusted for clarity and understanding of the message. The range of values that are accepted by the EMIC 2 is 75 (slowest) to 600 (fastest). The default speaking rate value is set to 200. The following is the example Arduino code to set the EMIC 2 TTS module's rate:

```
// Set speaking rate
  emicSerial.print("W100\n");
  while (emicSerial.read() != ':');   // wait for ':'
  character
```

- **Lx**: This sets the EMIC 2 TTS module's language.

  The final DecTalk speech command is selecting a language for the EMIC 2 TTS module. The default language is English (0), but two other languages can be selected for the EMIC 2 with the following code:

```
Castilian Spanish
Latin Spanish
Here is the example code:
  // Set language
  emicSerial.print("L2\n");
  while (emicSerial.read() != ':');   // wait for ':'
  character
```

  After uploading the modified sketch, the talking logic probe will speak the two phrases in Latin American Spanish.

Summary

In this chapter, a discussion on the theory and operation of a talking logic probe was covered. A basic talking logic probe design was shown along with assembly and testing instructions for building the speech synthesis electrical tester. Finally, technical details of how the EMIC 2 text-to-speech module operates were discussed using a series of hands-on experiments. The experiments showed how to set specific EMIC 2 DecTalk speech synthesizer engine feature/functions using Arduino code programming techniques.

In the next chapter, a hands-on investigation on using a web page as a **Human Machine Interface (HMI)** controller will be illustrated. As in the previous chapters, a series of lab experiments, followed by a final project will show how to build an Arduino-based HMI controller.

4
Human Machine Interface

The programmable DC motor controller with LCD project presented in *Chapter 2, Programmable DC Motor Controller with LCD*, is an example of a low design version of a **Human Machine Interface** (**HMI**). An HMI allows a human operator to operate a robot or intelligent machine using a control panel. The operator interacts with the DC motor HMI using a pair of electrical slide switches to turn the electromechanical machine on or off. The LCD provides a visual display of the DC motor's operating states of on or off. The disadvantage to an electromechanical-based HMI device is the mechanical wear of the electrical slide switch contacts. Deterioration of the slide switch contacts will not allow proper signal level detection (+5V or 0V) by the Arduino to operate the DC motor correctly. A solution to this problem is to use a software-based HMI design. An example of a software-based HMI design is shown in the following screenshot. In this example of HMI design, the electromechanical switches are replaced with graphical pushbuttons to eliminate mechanical contact wear.

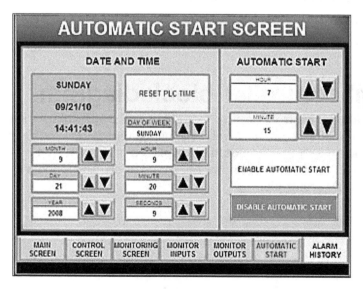

The words **human** and **operator** are used interchangeably in the industrial controls industry.

In this chapter, we will learn how to build a software HMI controller using a standard HTML-based web page. Also, a physical computing software package called **Breakout** will be introduced to the reader. This package will be used in building a simple HMI controller to operate a small DC motor and to monitor a pushbutton switch. A brief discussion on Breakout, the web page server, and the communications between these software elements will be provided in this chapter as well. This section will also provide instructions for building the HMI controller.

Parts list

The list of the parts required for HMI is as follows:

- Arduino Uno (one unit)
- Small DC motor (3V-6V DC operating range) (one unit)
- 1 kilo ohm resistor (black, brown, red, and gold) (one unit)
- 10 kilo ohm resistor (black, brown, orange, and gold) (one unit)
- Tactile pushbutton switch (one unit)
- 1N4001 or equivalent general purpose silicon diode (one unit)
- 2N3904 NPN transistor (one unit)
- Breadboard
- Wires

An HMI controller block diagram

An conceptual HMI controller can be thought of as a **graphical user interface** (**GUI**) with a DC motor attached to it. A web page using appropriate pushbuttons for turning on and off a DC motor would be displayed on the screen. Pressing either of the buttons will allow the DC motor to turn on or off. The following is a concept drawing of an HMI controller:

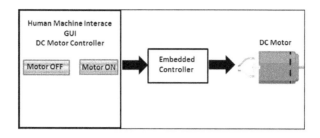

The HMI controller block diagram is an engineering development tool used to convey a complete product design using graphics. The block diagram also makes it easier to plan the breadboard for prototyping and testing of the HMI controller in a maker's workshop or laboratory bench. A final observation of the HMI controller block diagram is that the basic computer convention of inputs is on the left-hand side, the processor is located in the middle, and the outputs are placed on the right-hand side of the design layout. As shown, the desktop PC or notebook computer with the HMI GUI and pushbutton switch is on the left-hand side, the Arduino is located in the middle, and the DC motor is on the right-hand side of the block diagram. The desktop PC or notebook computer will provide serial binary data to the Arduino for processing DC motor control commands. The pushbutton switch is used to send an electrical signal to the Arduino. The pushbutton switch binary data will be displayed on the HMI GUI. This method of displaying data from an electromechanical component's status is another feature of an HMI GUI. The transfer of serial data is through a USB cable attached to both the computer and the Arduino. This left to right design method makes it easier to build the HMI controller and troubleshoot errors during the testing phase of the project.

 In addition to controlling electromechanical devices, HMI control panels can also monitor various industrial processes, such as temperature, pressure, and electromechanical component operating status.

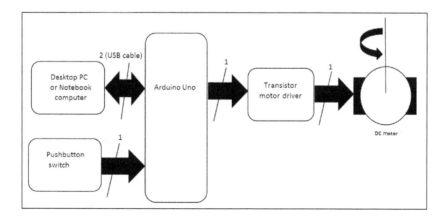

Testing the transistor motor driver

Typically, a microcontroller's digital pin can provide between 20-40 mA of current. This low current value is not able to turn on a small DC motor. The transistor's current gain h_{fe} is sufficient to the level from where it can directly operate the motor. A typical current gain value of 100 is appropriate for operating a small DC motor. This current gain slightly increases the microcontroller's digital pin value to a level capable of operating a small DC motor.

[Another name for a microcontroller's digital pin is **General Purpose Input-Output** or **GPIO**.]

The solderless breadboard diagram of the transistor motor driver is shown next. Note that the Arduino has a transistor wired to a digital pin to turn on a small DC motor. Pressing the pushbutton switch will turn on the transistor to operate the DC motor.

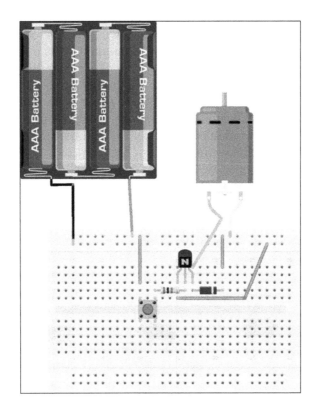

For an additional reference, the following is the transistor motor driver circuit schematic diagram:

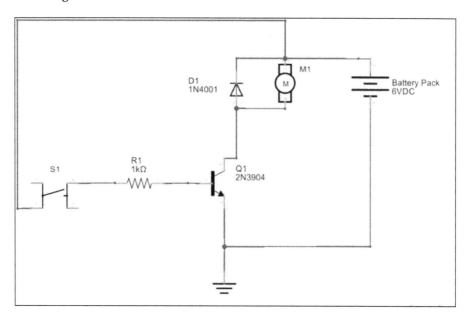

Pressing the pushbutton switch (**S1**) will allow the NPN transistor (**Q1**) to be properly biased by the base resistor (**R1**). With the transistor properly biased, the motor will turn on. The general purpose diode (**D1**) will prevent a reverse electrical voltage or transient signal from damaging the NPN transistor when the motor is turned off. Testing the transistor motor driver will remove any doubt about the circuit working properly while troubleshooting a faulty HMI controller.

Testing the pushbutton switch

The pushbutton switch operation is checked by taking a voltage measurement across the pull-down resistor. A digital or analog voltmeter is used to read the voltage across the pull-down resistor. Pressing the switch button will read approximately +5 V DC across the pull-down resistor. Releasing the switch button will have a 0 V DC reading on the voltmeter.

The following is the final DC motor controller breadboard diagram:

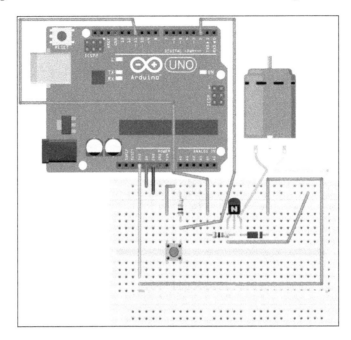

The circuit schematic diagram for the HMI DC motor controller is shown next and is provided as an additional project assembly tool:

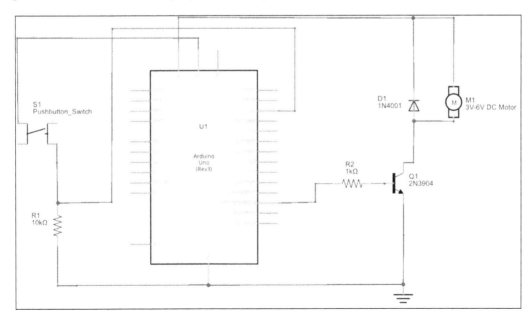

Making the web page physical

In order to interact with the Web, we need an electronic brain that can take its binary information and convert it into electrical signals to be used by the electromechanical and electronic devices, such as motors, switches, and LEDs. When the Arduino is receiving input data from electrical switches and sensors, the microcontroller will output a physical response using motors, relays, and LEDs. This technique of receiving signals from electronic sensors, electrical switches, and controlling electromechanical components and visual displays is an example of **physical computing**.

 The 3.3V source on the Arduino is used to prevent the small DC motor from heating up.

The electrical processing of input and output data occurring within the Arduino microcontroller requires a hardwired connection to the Web as follows:

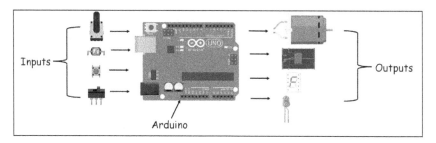

A local server will take the hardwired electrical signals from the Arduino and display them on a web page. In the other direction, motors, relays, and LEDs can be operated from the web page.

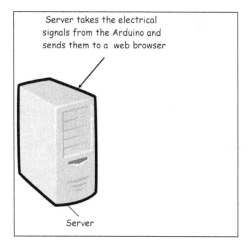

Physical computing deals with building devices that can sense and respond to their environment using software and hardware.

Now serving, the Arduino

Traditionally, a server waits for binary data from a web browser. The data is represented by images and sound. The server required for a physical computer uses electrical signals from an Arduino to connect to a web browser.

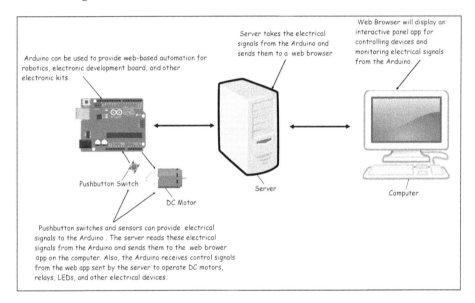

Getting into the real world using Breakout

Breakout is a software application used as an electronics prototyping tool to make web-based control and monitor panels. Electronic circuits, sensors, motors, LEDs, and other electrical and electromechanical devices can be controlled and monitored using this JavaScript-based tool. In a nutshell, Breakout is a virtual server that allows the Arduino to communicate with a web page, allowing the physical and virtual worlds to interact.

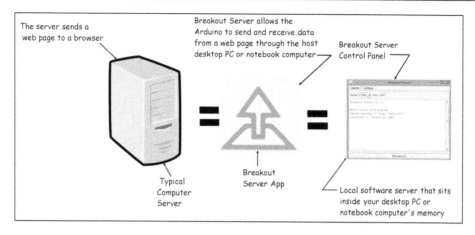

Pre-lab exercise

The objective to perform the pre-lab exercise is to have the Arduino communicate with a Breakout server.

The following are the steps to connect Arduino to Breakout server:

1. Connect the Arduino to the computer using a USB cable.

2. Download a copy of the Breakout server software from `http://breakoutjs.com`.

3. Install the Breakout server on your computer's hard drive by following the instructions for your computer's operating system (Windows, Linux, or Mac).

4. Download a copy of the Arduino **IDE** (**Integrated Development Environment**) from `http://arduino.cc`.

5. Within the Arduino IDE, upload the `StandardFirmata` sketch onto the Arduino. To do this, navigate to **File | Examples | Firmata | StandardFirmata**.

6. Open the Breakout server application you installed onto your computer in step 2.

7. Click on the **Connect** button on the Breakout server application.

8. The connection result to the Breakout server should be made as shown here:

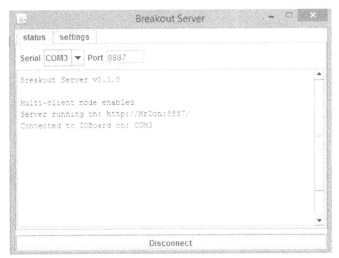

 Another alternative to obtaining the `Firmata` sketch is to navigate to the installed Breakout directory on your desktop PC or notebook computer's hard drive and upload the `AdvancedFirmata.ino` file to the Arduino—`Breakout/firmware/AdvancedFirmata/`.

Congratulations, your Arduino is now connected to a local server!

Setting up the Breakout file directory

The installation of the Breakout software is critical to the operation of the HMI DC motor controller. Once the folder has been unzipped to the specific drive location, the structure of the files is listed in a specific order, as shown in the following screenshot. Deviating from this file of `C:\Program Files (x86)\Breakout\example\ actuators layout` will cause improper operation of the HMI DC motor controller. The `Motor_Control_1220214.html` file, after it has been typed using a simple word processing application such as Notepad, can be placed into the actuator folder. The caution here is to assure that the files are in the order they were installed on your desktop PC or notebook computer for proper operation of the HMI motor controller.

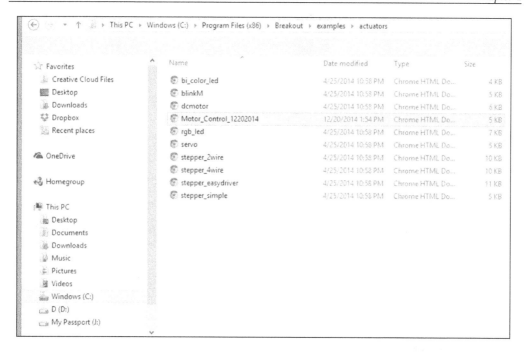

With the files properly stored to a specific drive location, they can easily be transported to any storage device. The reason for ease in transporting is the HTML format used in building the GUI control panel. The HTML files open in any web browser of your choice; this makes it convenient in terms of executing the HMI control panel on any OS platform.

The motor control HTML script

The HMI GUI is built from an HTML script that defines the look of the control panel. It also provides the physical layer in terms of the object definitions for operating the small DC motor and reading the status of the pushbutton switch. The following is the motor control HTML script:

```
    <!DOCTYPE html>
<html>
<head>

<meta charset=utf-8 />
```

```
<meta name="viewport" content="width=device-width, initial-
scale=1, maximum-scale=1">
<style type="text/css">
body {
    margin: 15px;
    font-family: sans-serif;
    font-size: 16px;
    line-height: 1.5em;
    color: #666;
}
h2 {
    padding-left: 0px;
    font-weight: normal;
    font-size: 28px;
    color: #00AEFF;
}
.ledBtns {
    padding: 10px;
    font-size: 22px;
    width: 130px;
    color: #00AEFF;
}
#state {
    color: #00AEFF;
    font-size: 22px;
    margin-bottom: 0;
}
.container {
    background-color: #f7f7f7;
    padding: 0 10px 20px 10px;
    border: 1px dotted #CCC;
    width: 270px;
    margin-top: 20px;
}
.spacer {
    margin-left: 5px;
}
#schematic {
    position: absolute;
    display: none;
    top: 65px;
    left: 15px;
```

```
}
#schematicBtn {
    margin-top: 20px;
}
</style>

<title>Motor Control Example</title>

<!-- The following (socket.io.js) is only required when using the
node_server -->
<script src="../../socket.io/socket.io.js"></script>
<script src="../../dist/Breakout.min.js"></script>
<script src="../libs/fastclick.min.js"></script>
<script src="../libs/jquery.min.js"></script>
<script type="text/javascript">
$(document).ready(function() {
  // Declare these variables so you don't have
  // to type the full namespace
  var IOBoard = BO.IOBoard;
  var IOBoardEvent = BO.IOBoardEvent;
  var LED = BO.io.LED;
  var Button = BO.io.Button;
  var ButtonEvent = BO.io.ButtonEvent;

  // Set to true to print debug messages to console
  BO.enableDebugging = true;

  // If you are not serving this file from the same computer
  // that the Arduino board is connected to, replace
  // window.location.hostname with the IP address or hostname
  // of the computer that the Arduino board is connected to.
  var host = window.location.hostname;
  // if the file is opened locally, set the host to "localhost"
  if (window.location.protocol.indexOf("file:") === 0) {
    host = "localhost";
  }

  // attach fastclick to avoid 300ms click delay on mobile
  devices
  FastClick.attach(document.body);

  var arduino = new IOBoard(host, 8887);

  // Variables
```

```
var led;
var button;
var $state = $('#state');

// Listen for the IOBoard READY event which indicates the
IOBoard
// is ready to send and receive data
arduino.addEventListener(IOBoardEvent.READY, onReady);

function onReady(event) {
  // Remove the event listener because it is no longer
  needed
  arduino.removeEventListener(IOBoardEvent.READY, onReady);

  // Create an LED object to interface with the LED wired
  // to the I/O board
  led = new LED(arduino, arduino.getDigitalPin(11));

  // Create a new Button object to interface with the
  physical
  // button wired to the I/O board
  button = new Button(arduino, arduino.getDigitalPin(2));

  // Listen for button press and release events
  button.addEventListener(ButtonEvent.PRESS, onPress);
  button.addEventListener(ButtonEvent.RELEASE, onRelease);

  // Use jQuery to get a reference to the buttons
  // and listen for click events
  $('#btnLeft').on('click', turnLedOff);
  $('#btnRight').on('click', turnLedOn);
}

function onPress(evt) {
  // Get a reference to the target which is the button that
  // triggered the event
  var btn = evt.target;
  // Display the state on the page
  $state.html("Button " + btn.pinNumber + " state: Pressed");
}

function onRelease(evt) {
  // Get a reference to the target which is the button that
  // triggered the event
```

```
      var btn = evt.target;
      // Display the state on the page
      $state.html("Button " + btn.pinNumber + " state: Released");
  }

  function turnLedOn(evt) {
      // Turn the LED on
      led.on();
  }

  function turnLedOff(evt) {
      // Turn the LED off
      led.off();
  }

</script>

</head>
<body>
  <h2>Motor Control Example</h2>
  <div class="container">
    <p><strong>Output:</strong> Use the buttons below to turn the
    motor on the breadboard ON or OFF.</p>
    <button id="btnLeft" class="ledBtns" type="button"> Motor
    Off</button>
    <span class="spacer"></span>
    <button id="btnRight" class="ledBtns" type="button">Motor
    On</button>
  </div>
  <div class="container">
    <p><strong>Input:</strong> Press the button on the breadboard
    to display the state below.</p>
    <p id="state">Display Button State</p>
  </div>
</body>
</html>
```

After typing the script into a simple word processor software package, such as Notepad or WordPad, save the document with an extension of .html in the actuator folder discussed earlier. Also, the file type should be a standard text document to allow the .html extension to be visible and recognized by any web browser once saved. Place this .html file into the actuators folder as discussed in the *Setting up the Breakout file directory* section of this chapter.

Execute the file by double-clicking on the name with the mouse. The contents of the file should open in a web browser displaying the HMI GUI, as shown in the following screenshot:

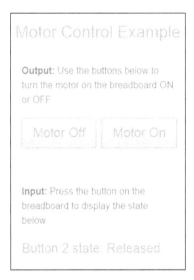

Clicking on the **Motor On** button with the mouse will turn on the transistor. The transistor will then operate the small DC motor wired to it. Clicking on the **Motor Off** button will turn off the motor. An important fact to keep in mind is that if the Breakout server is ever disconnected, clicking on the **Reconnect** button will re-establish the communication. Next, refresh the web page to make the buttons active for controlling the DC motor and reading the status of the pushbutton switch.

 The **Button 2 state** means the pushbutton switch is wired to digital pin 2 of the Arduino.

In reviewing the comments in the HTML code, the operation of the HMI DC motor controller can be discovered. For example, digital pin 10 can be used instead of pin 11 to control the motor by changing the following line of code:

```
// Create an LED object to interface with the LED wired
// to the I/O board
led = new LED(arduino, arduino.getDigitalPin(11));
```

The digital pin 10 is now assigned to the modified line of code:

```
led = new LED(arduino, arduino.getDigitalPin(10));
```

Likewise, the pushbutton switch can be assigned to a new digital pin by modifying this line of code:

```
// Create a new Button object to interface with the physical
// button wired to the I/O board
button = new Button(arduino, arduino.getDigitalPin(2));
```

The preceding line of code is modified to the new digital pin 5 assignment:

```
button = new Button(arduino, arduino.getDigitalPin(5));
```

There is a Wikipedia page on `Breakout.js` that explains in detail the programming objects API used in the HTML script for controlling and monitoring various electrical and electronic components like potentiometers, servo motors, and LED displays. The URL for the `Breakout.js` Wikipedia page is `http://breakoutjs.com/docs/`.

Summary

In this chapter, an HMI GUI was built using an HTML script to create a web page. The fundamentals of how a web page server works in the physical world using the `Breakout.js` software was discussed as well. The file structure for `Breakout.js` was explained using the installation directory given in this chapter as an example. Physical computing concepts were also discussed along with the examples of electromechanical and electronic components used in the construction of this interactive computation device. Finally, the HTML code for building and executing an HMI DC motor controller was given in this chapter along with the examples of how to change the Arduino for new motor and pushbutton switch digital pin assignments.

In *Chapter 5, IR (Infrared) Remote Control Tester*, a discussion on how to build an electronic testing device to check the operation of an IR remote control will be presented. Learning objectives in the chapter include IR detector operation, how to interface an IR detection circuit to the Arduino, and various visual displays to use for IR detection. As usual, mini lab testing procedures, Arduino code explanation, and the final project assembly will also be discussed in this chapter.

5
IR Remote Control Tester

Infrared (**IR**) is a wireless signal used in a variety of industrial, medical, and consumer electronics products. It is a part of the electromagnetic spectrum with its radiation and wavelength band between 770 nm and 1 mm. The wavelength of IR is longer compared to the visible light, which is in the range of 400-700 nm. In the industrial sector, it is known to be a heat radiation source and used in non-contact applications, such as object detection or level sensing. Also the energy radiated by infrared is not seen by the human eye, but it can be detected using special sensors. It's these visibility characteristics that make infrared useful in products such as thermal imaging cameras and object detection sensors known as **PIRs** (**Passive Infrared**).

In addition to object detection, IR is used in wireless control applications. The most common IR wireless control device used to operate electronic toys and consumer products is the handheld remote. A simple trick to check for a working IR handheld remote is to point the device at a smartphone's camera lens. Pressing any button on the handheld remote will display a small blue light on most smartphones' camera lenses.

Another alternative to the smartphone camera is to build an IR remote control tester using Arduino and a few off-the-shelf electronic components.

The smartphone camera lens method only works with smartphones that can detect IR signals. Therefore, the IR remote control tester is an alternative testing device to have on the workbench or lab desktop.

In this chapter, we will learn how to build an IR remote control tester using Arduino and **littleBits** electronic modules. Also the operation of IR detection will be discussed in this chapter. In addition to testing handheld remotes, this IR tester can be used as an electronic controller for operating DC motors and other small electromechanical components.

Parts list

The list of the parts required to build an IR remote control tester is as follows:

- Arduino Uno (one unit)
- littleBits power module (one unit)
- littleBits remote trigger module (one unit)
- littleBits proto module (one unit)
- littleBits number module (one unit)
- LCD – 16x2 (one unit)
- Breadboard
- Wires

An IR remote control tester block diagram

An IR remote control tester can be thought of as an electronic sensor with a visual display. A text message of the IR signal being present or absent will be displayed on a visual screen. Placing an IR handheld remote control in front of an electronic sensor and pressing a button will show the presence of an infrared signal on the visual display. Here is a concept drawing of an IR remote control tester:

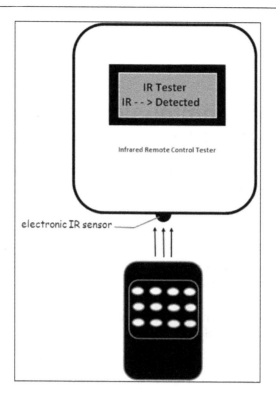

The IR remote control tester block diagram is an engineering development tool used to convey a complete product design using graphics. The block diagram also makes it easier to plan the breadboard for prototyping and testing of the IR remote control tester in a maker's workshop or laboratory bench. A final observation of the IR remote control tester block diagram is that the basic computer convention of inputs is on the left-hand side, the processor is located in the middle, and the outputs are placed on the right-hand side of the design layout. As shown in the following diagram, the IR receiver module is on the left-hand side, the Arduino is located in the middle, and the **Liquid Crystal Display** (**LCD**) is shown on the right-hand side. The IR receiver will provide the on/off binary data to the Arduino for processing and LCD control.

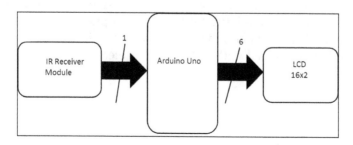

The IR receiver module receives serial binary data from a handheld remote and converts it to an on/off control signal.

IR signals and communication protocols

The majority of today's IR handheld remote controls, for example, consumer electronics devices such as home entertainment equipment, use small data rates for transmitting wireless control signals. A typical data rate used for an IR remote control is about 2400 **bps (bits per second)**. Transmitted signals from an IR handheld remote are switched at a high rate of speed, allowing the control distance to be typically up to 20 feet.

In order to achieve a significant transmission distance, the control signal is modulated. **Modulation** is a technique where the control signal (data) is placed inside of a high frequency carrier wave. A common modulation value for an IR remote control signal is 38 kHz (kilohertz). A control signal is a series of binary pulses that can be decoded using a microcontroller. The following is the image of a typical IR remote control signal:

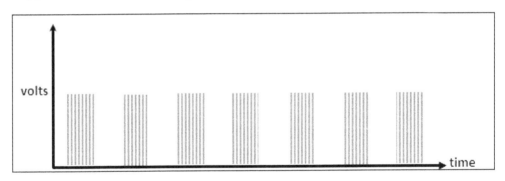

To ensure that all the IR remote control manufacturers provide proper control signals that will not interfere with other electronic devices, the **Infrared Data Association (IrDA)** was established in 1993. The IrDA provides design rules (protocols) to help facilitate secure data transfers of all wireless transmissions from IR remote controls. Also, **Line of Sight (LOS)** techniques for IR wireless optical communications are managed by the IrDA. As a maker, the infrared communication protocols for our IR remote control are managed within the littleBits electronic remote trigger-receiving module.

littleBits electronic modules

To make the IR remote control prototyping experience more enjoyable, littleBits electronics modules are being introduced in this chapter. The littleBits are colorful electronic modules that provide specific electrical functions for consumer devices and gadgets. By connecting the modules in unique and creative ways, interesting, innovative electronic devices and gadgets can be built.

There are three metal pins that allow the littleBits electronics modules to work properly. They consist of the following signal names:

- vcc (+5 V power supply)
- sig (electrical signal)
- gnd (electrical ground)

The littleBits electronic modules include plastic **bitSnaps** that have three metal pins for providing the electrical connections. The bitSnap electrical connections are shown in the following figure.

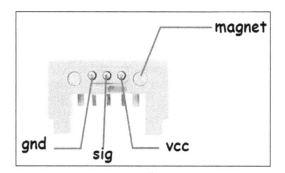

Check out the littleBits electronics modules at
http://littlebits.cc/.

To reduce error in building a gadget or device, small magnets are placed inside the littleBits bitSnaps. The electronic modules will connect to each other when properly attached. Incorrect electronic module connections will repel each other. The littleBits modules given in the *Parts list* section of this chapter will be used to construct the IR receiver module in the block diagram.

The following are the examples of littleBits for the IR remote control tester prototype:

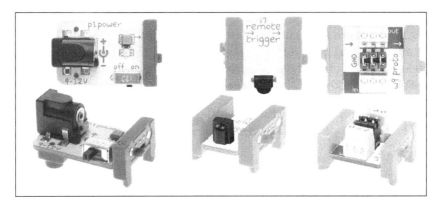

To identify the four basic types/functions of littleBits electronic modules, the following color code scheme is used: blue (power), orange (wire), pink (input), and green (output). As seen in the previous image, these are signified by the bitSnaps of the modules.

Wiring the IR receiver module

The IR receiver module is an electronic sensor able to detect infrared signals. To rapidly build the actual remote control tester, a littleBits electronic module will be used. The remote trigger module has an IR sensor (TSOP382) and a pre-amplifier circuit soldered onto a small **printed circuit board** (**PCB**). The TSOP382 sensor is an IR receiver circuit packaged inside a small three lead epoxy component. While detecting an IR signal, the TSOP382 will generate a series of binary pulses, as shown in the *IR signals and communication protocols* section of this chapter. The following is the image of the littleBits remote trigger module and TSOP382 IR sensor:

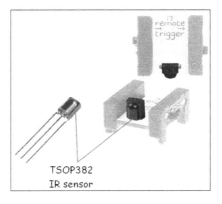

In addition to the TSOP382 IR sensor, the remote trigger module consists of several **operational amplifiers (op-amps)** and transistors. The remote trigger op-amps are used for IR signal conditioning and output buffering of the binary data. The transistors provide small gain amplification and output switching for the remote trigger module. The complete circuit schematic diagram of the remote trigger module can be viewed at `http://littlebits.cc/bits/remote-trigger`. It can be found under the **Extra Materials** heading on the web page.

As discussed in the previous chapters, the block diagram is the engineering tool used to build and test Arduino-based electronics. As shown in the following revised block diagram, all the components needed for the IR remote control tester are connected to form an interactive block diagram; the littleBits modules share the three electrical signals, allowing the connected PCBs to function correctly:

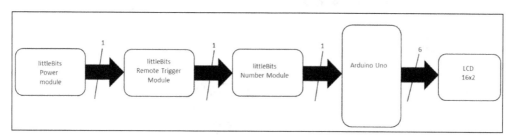

One of the key innovations of the littleBits electronic modules is the shared power rails and distributed signal using the magnetic-based electrical connector system. The littleBits power module provides a +5V DC voltage to the other electronic modules. The +5V DC voltage is present through the bitSnaps. The remote trigger provides a control signal to the number module and the Arduino when an IR signal is present. The control signal value is displayed on the number module. The Arduino receives the control signal and displays a message on the LCD.

To test the remote trigger sensor, the littleBits power and number modules will be connected as follows:

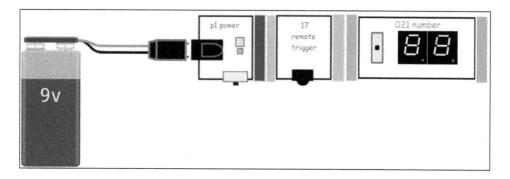

Pressing any button on an IR remote control, placed in front of the remote trigger module, will display a numeric value on the number display, as shown in the next diagram. The numeric value can either display the voltage level applied (voltmeter mode) to the number module's sig pin or its equivalent **analog to digital converter (ADC)** count. The small slide switch soldered to the PCB will allow selection between these two numeric display modes. The maximum ADC count value that can be displayed is **99** with an equivalent voltage value of 5.0. Also, the numeric value displayed will vary based on the 9 V battery's output voltage level used with the littleBits power module as well. The following is the example of the slightly less than 9 V battery output voltage:

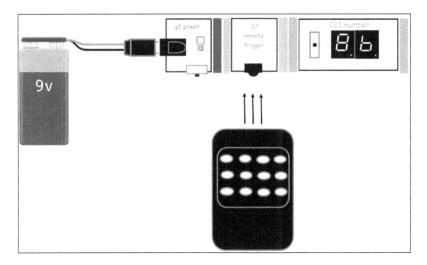

To illustrate the technology elegance of the littleBits power module, a circuit schematic diagram for your reference is shown here:

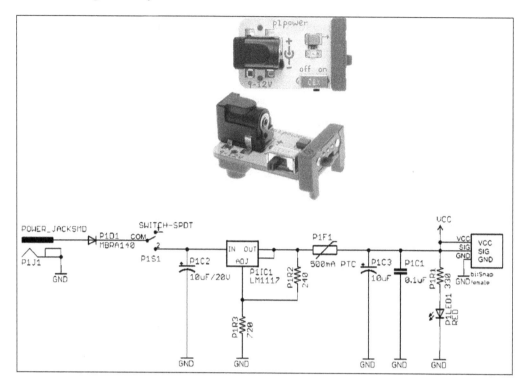

 Check out the littleBits electronics modules at the URL address http://littlebits.cc/.

The slide switch on the number module can be positioned to volts to see the output voltage produced by the remote trigger when activated by an IR remote control. A typical value of 4.2 V will be seen on the number module when pressing a button on an IR remote control. The littleBits number module is shown in the following image:

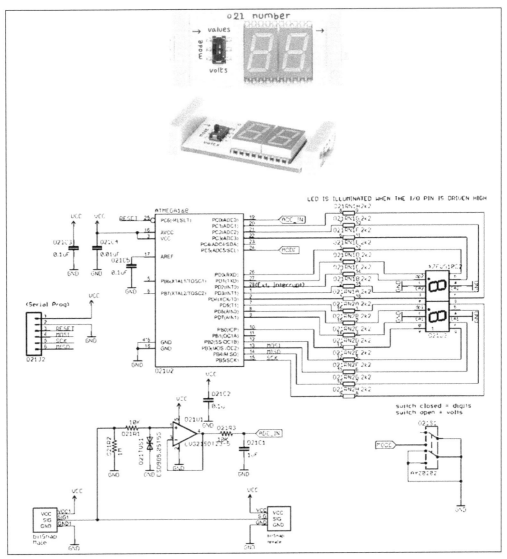

The littleBits number module

The circuit schematic diagram can be viewed at http://littlebits.cc/bits/ number. It can be found under the **Extra Materials** heading on the web page.

Wiring the Arduino and the LCD

With the littleBits remote trigger and number electronic modules working correctly, it's time to complete the tester. The next two components needed for the final prototype build are the Arduino and the LCD. To connect the Arduino to the littleBits electronic modules, a special interface connector known as **proto module** is needed. The proto module, shown next, has two small terminal blocks and three jumper pins soldered on a mini PCB. The three electrical signals (vcc, sig, gnd) are accessible from the terminal blocks.

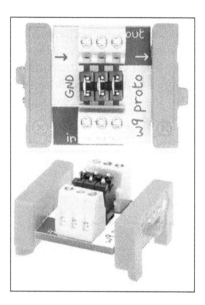

The location of the proto module's electrical signals is shown here:

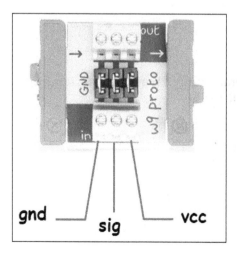

The placement of the proto module with respect to the other littleBits devices should be between the remote trigger and number modules, as shown here:

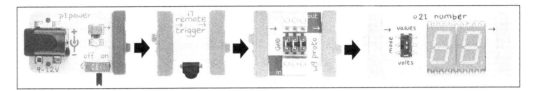

The location of the proto module to the remote trigger is set to obtain the electrical signal produced when the electronic sensor is activated by an IR remote control. The electrical signal produced by the remote trigger will be wired to the Arduino with an LCD. As shown in the IR remote tester concept drawing, the LCD will provide a status message on the absence or presence of a detected infrared signal. The following is the image of the electrical wiring for the IR remote tester, based on the arrangement of the littleBits electronics module discussed:

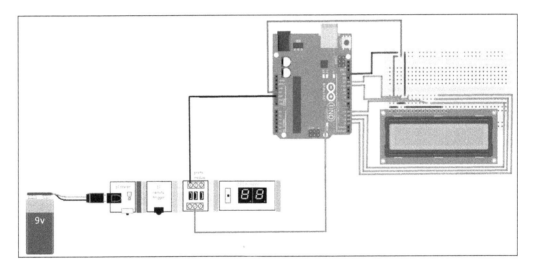

The 16x2 LCD has several required wiring connections to make it operational. The required LCD pins are listed here:

- Pin 1 (Vss [gnd])
- Pin 2 (Vdd [+V])
- Pin 3 (Contrast [gnd])
- Pin 4 (**Register Select[RS]**)

- Pin 5 (**Read/Write [R/W]**)
- Pin 6 (Enable)
- Pin 11 (D4: [8 bit data bus])
- Pin 12 (D5: [8 bit data bus])
- Pin 13 (D6: [8 bit data bus])
- Pin 14 (D7: [8 bit data bus])
- Pin 15 (LED – [gnd])
- Pin 16 (LED + [+V])

Some LCD manufacturers make 14 pin LCDs only. As seen in the LCD pinout list, the first 14 pins provide functionality for the solid state display when correctly wired to the DC power supply rails (+5 V, gnd) and the Arduino microcontroller. As an electrical wiring reference for the LCD pinout, the following is the circuit schematic diagram showing this solid state display wire to the Arduino; the littleBits proto module's electrical signal pin (sig) is wired to digital pin 6 (D6) of the Arduino:

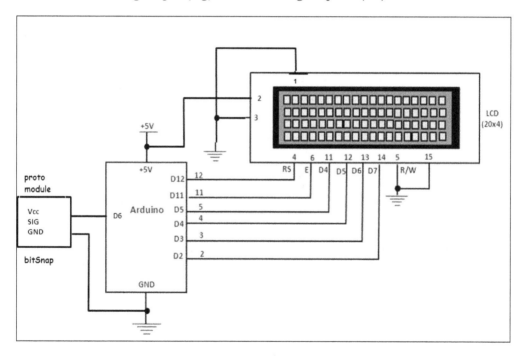

The LCD displays the status message *IR à Detected* when the Arduino receives an electrical signal from the littleBits remote trigger. The littleBits proto module provides an electrical interface for the Arduino to obtain the remote trigger's control signal activated by an IR remote control. The following is a circuit schematic diagram showing the electrical interface of the littleBits remote trigger to the Arduino using the proto module:

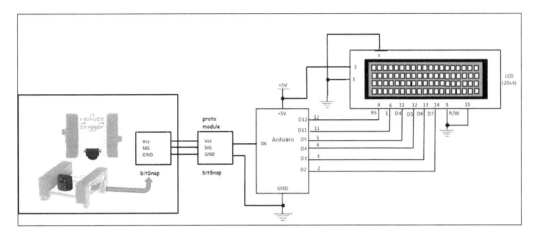

 The proto module can be purchased separately or with supporting electrical interfacing components sold with the littleBits **Hardware Development Kit** (**HDK**) bundle. The URL address for the littleBits HDK is `http://littlebits.cc/kits/hdk`.

IR Tester code

With the littleBits electronic modules working correctly, building IR Tester code is the next project step. Basic software requirements for the IR Tester codes are:

- Pressing any button on the IR remote control will display the status message *IR à Detected* on the LCD

- No button pressed on the IR remote control will display the status message *IR not Detected* on the LCD

To visually see the logic flow of these basic software requirements in action, an interactive flowchart is shown here:

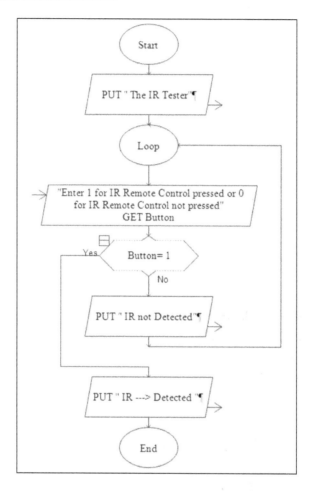

The interactive flowchart is built using a free download software package called **Raptor**. By running the Raptor interactive flowchart, each symbol will be executed in real time. The executed symbol is highlighted in green. Running the Raptor interactive flowchart when an IR remote control button is pressed is as follows:

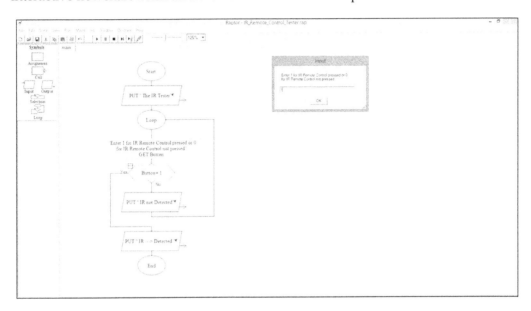

The LCD text message displayed on the Master Console is shown here:

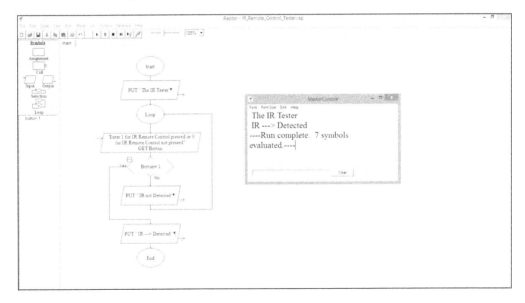

The Raptor interactive flowchart software can be downloaded from `http://raptor.martincarlisle.com/`.

The pseudocode used in the Raptor interactive flowchart can be turned into Arduino code, as shown here. Two versions of code can be built for displaying status messages on a serial monitor and LCD. The serial monitor code version is as follows:

```
#include<LiquidCrystal.h>; // LCD library
int IRout = 7; // Arduino digital pin used to detect littleBits IR
signal
int IRoutStatus = 0; // variable used to track detection event of
Arduino digital pin
LiquidCrystal lcd(12, 11, 5, 4, 3, 2);// Arduino pins wired to
16x2 LCD display

void setup() {
  pinMode (IRout, OUTPUT);// configuring IRout pin as an output
  Serial.begin(9600); // configuring USB port for serial
  communication
  Serial.println(" IR Tester");//display tester title on the
  Serial Monitor
  Serial.println(""); //print a blank line

  lcd.begin(16,2);// establishing LCD as a 16x2 opto electronic
  device
  lcd.setCursor(0,0);// setting Column, Row for LCD text (starting
  point)
  lcd.print("   IR Tester"); // to center tester name on LCD, 4
  spaces required
}

void loop() {
  IRoutStatus = digitalRead(IRout); // read digital value of IRout
  output pin
  if (IRoutStatus == HIGH){ // if status of IRoutStatus is HIGH...
    Serial.print("IR --> Detected"); // print text on Serial
    Monitor
    Serial.println("");// print a space for carriage return on
    Serial Monitor
    lcd.setCursor(0,1); // set Column, Row LCD location
    lcd.print("IR --> Detected");// print text on LCD
    delay(1000);//provide 1 sec delay between prints on serial
    monitor
  }
```

```
else{
    Serial.print("IR not Detected");// if status of IRoutStatus in
    not HIGH...
    Serial.println("");// print a space for carriage return on
    Serial Monitor
    lcd.setCursor(0,1);// set Column, Row LCD location
    lcd.print("IR not Detected");// print text on LCD
    delay(1000);//provide 1 sec delay between prints on serial
    monitor
    }
}
```

Uploading the code to the Arduino displays the following results on the serial monitor:

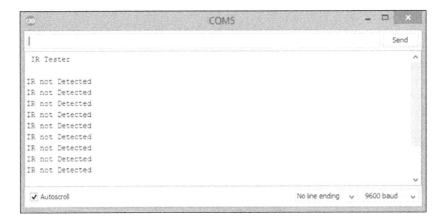

Pressing any button on the IR remote control, displays the following status message on the serial monitor:

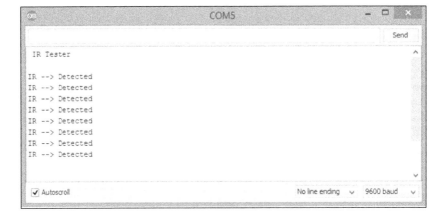

You can modify the code to allow the same serial monitor status messages to be displayed on the 16x2 LCD:

```
#include<LiquidCrystal.h>; // LCD library
int IRout = 7; // Arduino digital pin used to detect littleBits IR
signal
int IRoutStatus = 0; // variable used to track detection event of
Arduino digital pin
#include<LiquidCrystal.h>; // LCD library
LiquidCrystal lcd(12, 11, 5, 4, 3, 2);// Arduino pins wired to
16x2 LCD display

void setup() {
  pinMode (IRout, OUTPUT);// configuring IRout pin as an output

  lcd.begin(16,2);// establishing LCD as a 16x2 opto electronic
  device
  lcd.setCursor(0,0);// setting Column, Row for LCD text (starting
  point)
  lcd.print("   IR Tester"); // to center tester name on LCD, 4
  spaces required
}

void loop() {
  IRoutStatus = digitalRead(IRout); // read digital value of IRout
  output pin
  if (IRoutStatus == HIGH){ // if status of IRoutStatus is HIGH...
    lcd.setCursor(0,1); // set Column, Row LCD location
    lcd.print("IR --> Detected");// print text on LCD

  }
else{
    lcd.setCursor(0,1);// set Column, Row LCD location
    lcd.print("IR not Detected");// print text on LCD
  }
}
```

Uploading the new code to the Arduino will allow the two status messages for IR remote control testing to be displayed on the LCD. Congratulations on building the IR remote control tester prototype!

 This IR remote control tester can be used to control a small transistor motor driver circuit using the littleBits HDK quite easily.

Summary

In this chapter, an IR remote control tester prototype was built using an Arduino, a 16x2 LCD, and a littleBits remote trigger sensor module. A discussion on IR signals and communication protocols was provided in this chapter. The littleBits electronic modules were introduced in this section as a rapid prototyping tool for building electronic devices easily. Another design tool for code development, called Raptor, was also introduced. Raptor is a free software download that allows building interactive flowcharts. The interactive flowchart for the IR remote tester's pseudo programming language was converted to the Arduino code. Two versions of code were built using the serial monitor and the 16x2 LCD to display status messages of detected infrared signals from an IR remote control. Wiring diagrams and circuit schematic diagrams for the electronics were provided and discussed in this chapter.

In *Chapter 6, A Simple Chat Device with LCD*, you will learn how to send text messages to an Arduino using **Bluetooth Low Energy** (**BLE**) device and an Android smartphone. Learning objectives in the chapter include introduction to the RedBearLabs Arduino shield and the Nordic nRF0001 BLE IC, displaying text messages on an LCD, and pairing the RedBearLabs shield with an Android smartphone

As usual, mini lab testing procedures, Arduino code explanation, and the final project assembly will be discussed in the next chapter.

6

A Simple Chat Device
with LCD

A **teleprinter** is an electromechanical typewriter that can be used to send and receive typed messages. The teleprinter can be arranged so that the typed messages can originate from point to point or point to multiple point networks using a variety of communication channels. Teleprinters were used to provide a simple user interface to interact with mainframe computers and minicomputers from the late 1930s to the early 1970s. They sent typed data to the mainframe computer and received printed responses from it.

Teletype (TTY) is another word used to describe a teleprinter.

The next stage in the electronic technology evolution of the teleprinter was the **pager**. A pager (commonly known as a beeper) is a wireless electronic communication device that receives and displays alphanumeric text messages. There are two commonly known pagers used for electronic wireless communication—one-way pagers and two-way pagers. One-way pagers can only receive messages, while two-way devices can provide bidirectional communications. The pager is able to send and receive data from computers or RF base stations using a special communication method called **Telelocator Alphanumeric Protocol (TAP)**. This method involves using modem connections wired directly to a paging network. TAP's data structure consists of a 10-bit **American Standard Coded Information Interchange (ASCII)** format for representing the intended transmitted text message, as shown here:

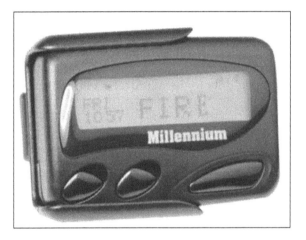

 TAP is an industry standard protocol capable of sending short text messages.

In this chapter, we will learn how to build equivalent teleprinter and pager messaging products know as **Simple Chat devices**. As in the previous chapters, the Arduino Uno will serve as the main processing device. The littleBits number module and an LCD along with a RedBearLab **Bluetooth Low Energy (BLE)** shield will provide visual and wireless communication features to enhance the user interface experience. Also, an Android smartphone and the RedBearLab BLE mobile app will be used to send and receive messages from a USB connected to a desktop PC or notebook computer.

The operation of the Nordic nRF8001 BLE IC will also be discussed in this chapter. In addition, a littleBits buzzer module can be added to the Simple Chat device (teleprinter) to provide an audible alert feature, triggered when a short text message is sent by a desktop PC or notebook computer.

Parts list

The following is the list of parts required for building a Simple Chat device with LCD:

- Arduino Uno (one unit)
- littleBits proto module (one unit)
- littleBits number module (one unit)
- LCD—16x2 (one unit)
- Breadboard
- Wires

A Simple Chat device block diagram

A Simple Chat device can be thought of as an electronic teleprinter. A text message can be sent to a visual display, such as the Arduino serial monitor or a typical LCD. The method of physically connecting to these visual displays can be accomplished using hardwired serial or wireless communications techniques. Creating the desired text message to send to the Simple Chat device will be accomplished using a typical notebook computer or desktop PC keyboard. These items are the basic physical requirements of our Simple Chat device. The concept drawings of two Simple Chat devices are as follows:

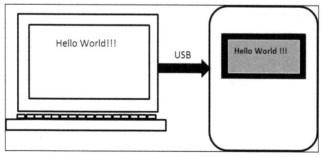

Serial communication-based Simple Chat device

Concept drawing of Simple Chat devices using BLE is as follows:

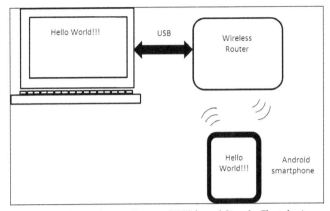

Wireless Bluetooth Low Energy (BLE) based Simple Chat device

Simple Chat device block diagrams are engineering development tools used to convey complete product designs using graphics. The block diagrams also make it easier to plan the breadboard for prototyping and testing of the Simple Chat devices in a maker's workshop or laboratory bench. A final observation of the Simple Chat device's block diagrams is that the basic computer convention of inputs is on the left-hand side, the processor is located in the middle, and the outputs are placed on the right-hand side of the design layout. As shown, the desktop PC or notebook computer is on the left-hand side, the Arduino is located in the middle, and the LCD with littleBits number module are on the right-hand side of the block diagram. The littleBits two digit LED display module will show the number 99 while the LCD is receiving a text message. The USB cable connected between the Arduino Uno and computer allows text message data to be entered into the serial monitor using a keyboard. The Android smartphone also allows text message data to be sent to the Arduino Uno using a BLE shield. Also, the Android smartphone can receive text message from the Arduino Uno using the serial monitor.

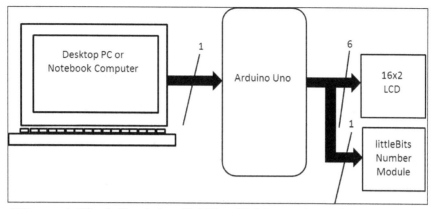

USB connected Simple Chat device with LCD and littleBits number module block diagram

Block diagram of Simple Chat devices using wireless technology is as follows:

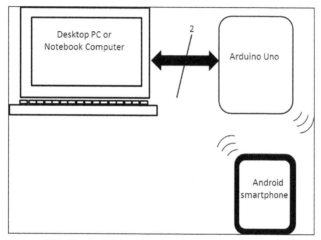

Wireless based Simple Chat device block diagram

 Wireless is the marketing word for **Radio Frequency (RF)**.

Building a serial-based Simple Chat device

The serial-based Simple Chat device uses a hardwired interface communication method of sending text message data. The USB cable provides a transmit wire that carries the binary equivalent character data code (ASCII) from a desktop PC or notebook computer to the Arduino Uno. The Arduino Uno code will convert the ASCII code into the appropriate binary data capable of displaying the text message onto the LCD.

Building the serial-based Simple Chat device is quite simple and requires only a few off-the-shelf electronic components. As discussed earlier, building the Simple Chat device is aided by using a block diagram. The prototype serial-based Simple Chat device using a solderless breadboard and a few off-the-shelf electronic components is shown here. As an additional reference, the circuit schematic diagram is also provided:

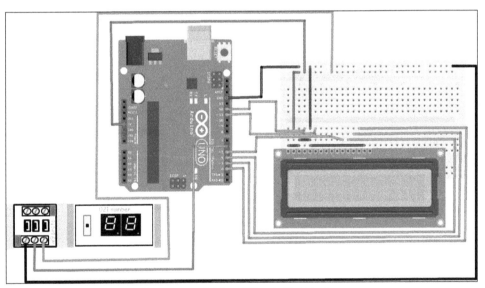

Prototype serial-based Simple Chat device

Circuit schematic diagram of Prototype serial-based Simple Chat device is shown as follows:

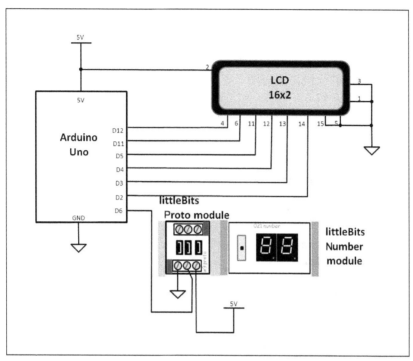

Circuit schematic diagram of Prototype serial-based Simple Chat device

An initial test of the LCD wiring consists of uploading the Hello World code that is part of the Arduino projects packaged with the IDE. To find the Hello World code within the IDE, the **Example | LiquidCrystal | Hello World.** directory tree is provided. If the LCD doesn't display the **hello world!** text message and the time in seconds, remove the USB connector from the Arduino. Correct errors using the wiring and circuit schematic diagrams and repeat the code test.

Serial-based Simple Chat device code

With the prototype hardware running properly, based on the initial testing, the final code can be uploaded to the Arduino. The following is the serial-based Simple Chat device Arduino code with LCD:

```
// include the library code:
#include <LiquidCrystal.h>
// Arduino digital pin number assigned to the littleBits number
modules
```

```
    int littleBits_Number = 6;


    // initialize the library with the numbers of the interface pins
    LiquidCrystal lcd(12, 11, 5, 4, 3, 2);

    void setup() {
      // set up the LCD's number of columns and rows:
      lcd.begin(16, 2);
      // initialize the serial communications:
      Serial.begin(9600);
      //littleBits Number module digital pin configured as an output
      pinMode(littleBits_Number, OUTPUT);
      //digitalWrite(littleBits_Number, LOW);
    }

    void loop(){
      // when characters arrive over the serial port...
      if (Serial.available()) {

        // wait a bit for the entire message to arrive
        delay(100);
        // clear the screen
        lcd.clear();
        // read all the available characters
        while (Serial.available() > 0) {
          // display each character to the LCD
          lcd.write(Serial.read());
          // turn on littleBits Number module during text message
          transmission
          digitalWrite(littleBits_Number, HIGH);
          // transmit text message and display 99 value on littleBits
          number module for 100ms
          delay(100);
          //stop text message transmission and display 00 value on
          littleBits number module
          digitalWrite(littleBits_Number, LOW);
        }
      }
    }
```

After the code has been uploaded to the Arduino Uno, open the serial monitor and type in the **I'm Arduino** text message, as shown in the following screenshot:

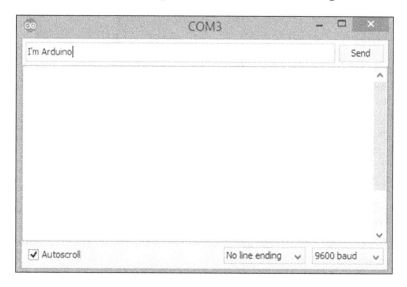

As the text message is being typed across the LCD, the littleBits number module displays the value of **99** as a transmission data validation indicator. The final result of the text message displayed on the LCD is shown here:

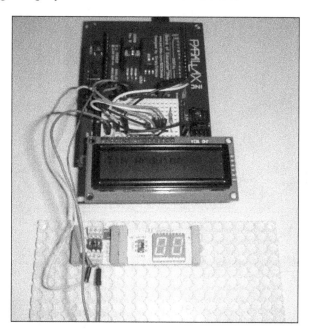

The magic behind the prototype serial Simple Chat device with LCD is based on the following lines of Arduino code:

```
// when characters arrive over the serial port...
   if (Serial.available()) {

      // wait a bit for the entire message to arrive
      delay(100);
      // clear the screen
      lcd.clear();
      // read all the available characters
      while (Serial.available() > 0) {
         // display each character to the LCD
         lcd.write(Serial.read());
```

The <Serial.h> library allows the text message to be converted into its equivalent ASCII code. This 8-bit binary code is then sent along the transmit wire of the USB cable to the Arduino Uno from the attached desktop PC or notebook computer. The <LiquidCrystal.h> library will change the 8-bit ASCII code to the appropriate binary data of the LCD internal register circuits. It's here the internal register circuits will command the 16 pixel blocks to display the typed text message from the serial monitor onto the LCD component. For reference, the following is a typical LCD block diagram:

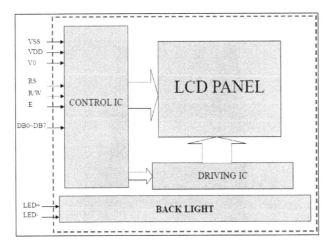

The Control IC has the internal register circuits for operating the 16 pixel blocks of the LCD.

The Nordic nRF8001 BLE IC

The final Simple Chat device, which will be explored later in the chapter, uses a wireless method of sending and receiving text messages. This electronics communication technique is dependent upon a single integrated circuit called the **Nordic nRF8001**. This IC is a complete BLE radio link circuit manufactured inside the QFN32 package. The BLE-compliant radio link provides the **physical layer (PHY)** and slave mode controller that allows the IC to connect to the real world. The PHY is accomplished through a serial interface, called **Application Serial Interface (ASI)**, which is easily configured by a connected microcontroller and software. The following is the pinout diagram of the nRF8001 BLE QFN32 IC:

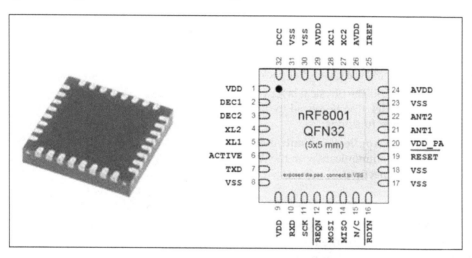

The nRF8001 BLE IC internal architecture consists of a low energy stack that manages the link controller and host stack. These two subcircuit blocks provide the main PHY for serial communication interfacing to a microcontroller. It also includes additional analog subcircuits needed for its BLE operation, such as power management and several oscillator options for assisting in wireless application functions.

The nRF8001 BLE IC internal architecture block diagram is as follows:

QFN32 is a Quad Flat pack 32 pin integrated circuit package.

PHY is the physical layer of the **Open Systems Interconnections** (**OSI**) model. It's referenced to the circuitry required to implement physical layer functions.

The **Application Controller Interface** (**ACI**) allows the nRF8001 BLE to communicate with the Arduino Uno using a hardwired serial data connection.

The unique feature that makes the nRF8001 BLE IC appealing to Maker wireless controls and monitoring projects is its low current requirements. The maximum current or peak draw for the nRF8001 BLE IC is as low as 12.5 mA (milliamperes) with an average current down to 9 uA (microamperes). With low current draw capabilities, the nRF8001 BLE IC enables battery lifetimes of months to years from a single coin cell (3.3 V).

The RedBearLab BLE shield

To aid the development of the Arduino Uno BLE applications, RedBearLab has created a user-friendly shield. Like traditional shields, the BLE unit snaps on the top of the Arduino.

The Arduino Uno is attached to the desktop PC or notebook computer using a USB cable, as shown in the following block diagram:

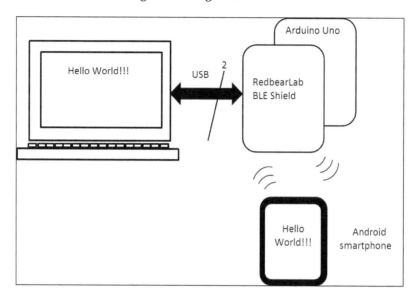

The text message is typed into the serial monitor and sent to an Android phone. As shown in the block diagram, both transmit and receive wires of the USB cable are used to allow bidirectional text messaging events to occur between the desktop PC or notebook computer and the Android smartphone. The following is an actual image of the RedBearLab BLE shield:

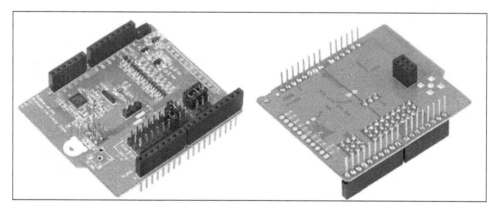

Attaching the RedBearLab BLE shield on the top of the Arduino Uno makes for a compact wireless device package:

Upon making the USB connection between the desktop PC or notebook computer and the Arduino Uno, a small blue LED power indicator is illuminated. The next phase of the BLE Simple Chat device project is to install the nRF8001 software library within the Arduino IDE.

Installing the RBL_nRF8001 library

To add the nRF8001 library to the Arduino IDE, go to the RedBearLab website at `http://redbearlab.com`. On the home page, click on the **+BLE Products** to open up a drop-down list and select the **BLE Shield** option. Scroll down the web page until the **Resources** section is reached and select the `RedBearLab nRF8001 library` link:

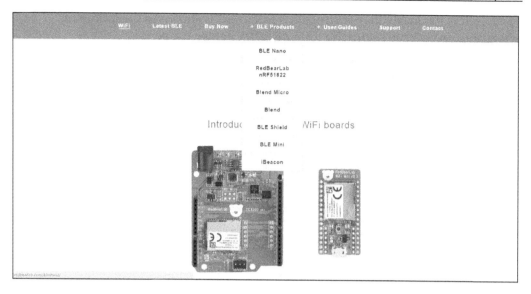

The GitHub web page with the Arduino nRF8001 library and software will appear as follows:

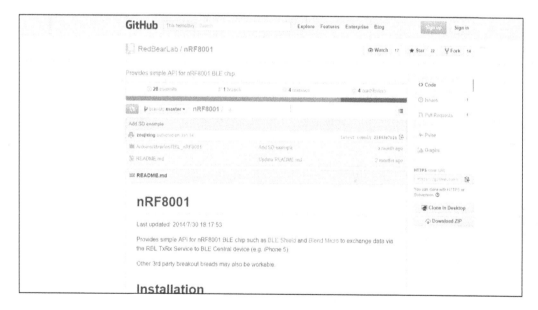

Next, follow the installation instructions carefully, as shown in the following screenshot, in order to set up the RBL_nRF8001 library properly:

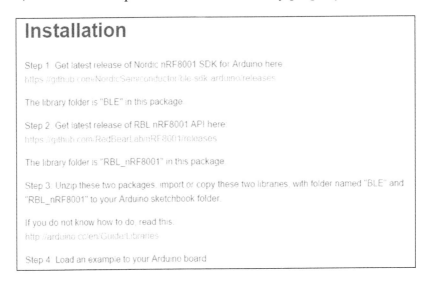

 Additional installation instructions can be found at http://redbearlab.com/getting-started-bleshield.

After installing the RBL_nRF8001 library files onto the target hard drive, the file directory will appear as follows:

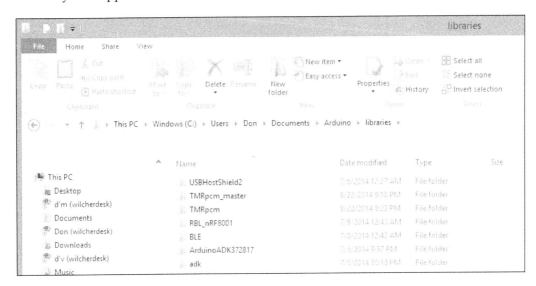

If the installation instructions are not followed explicitly, the BLEControllerSketch code will not compile correctly for the Arduino Uno.

After correctly installing the RBL_nRF8001 library, the example code files (shown here) will be available within the Arduino IDE:

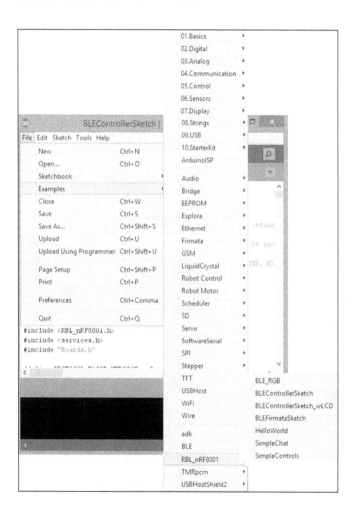

Uploading the BLEControllerSketch code to the Arduino Uno

With the RBL_nRF8001 library installed in the Arduino IDE, the next development step for the BLE-based Simple Chat device is to upload the wireless code to the Arduino Uno. RedBearLab has written a wireless application code that allows the Arduino to communicate with Android smartphones. The BLEControllerSketch is the target application code that will allow an Android smartphone to send and receive text messages from the Arduino IDE serial monitor. The code for the BLEControllerSketch is shown as follows:

```
#include <Servo.h>
#include <SPI.h>
#include <boards.h>
#include <RBL_nRF8001.h>
#include <services.h>
#include "Boards.h"

#define PROTOCOL_MAJOR_VERSION   0 //
#define PROTOCOL_MINOR_VERSION   0 //
#define PROTOCOL_BUGFIX_VERSION  2 // bugfix

#define PIN_CAPABILITY_NONE      0x00
#define PIN_CAPABILITY_DIGITAL   0x01
#define PIN_CAPABILITY_ANALOG    0x02
#define PIN_CAPABILITY_PWM       0x04
#define PIN_CAPABILITY_SERVO     0x08
#define PIN_CAPABILITY_I2C       0x10

// pin modes
//#define INPUT                 0x00 // defined in wiring.h
//#define OUTPUT                0x01 // defined in wiring.h
#define ANALOG                 0x02 // analog pin in analogInput
mode
#define PWM                    0x03 // digital pin in PWM output
mode
#define SERVO                  0x04 // digital pin in Servo
output mode

byte pin_mode[TOTAL_PINS];
byte pin_state[TOTAL_PINS];
byte pin_pwm[TOTAL_PINS];
```

```
byte pin_servo[TOTAL_PINS];

Servo servos[MAX_SERVOS];

void setup()
{
  Serial.begin(57600);
  Serial.println("BLE Arduino Slave");

  /* Default all to digital input */
  for (int pin = 0; pin < TOTAL_PINS; pin++)
  {
    // Set pin to input with internal pull up
    pinMode(pin, INPUT);
    digitalWrite(pin, HIGH);

    // Save pin mode and state
    pin_mode[pin] = INPUT;
    pin_state[pin] = LOW;
  }

  // Default pins set to 9 and 8 for REQN and RDYN
  // Set your REQN and RDYN here before ble_begin() if you need
  //ble_set_pins(3, 2);

  // Set your BLE Shield name here, max. length 10
  //ble_set_name("My Name");

  // Init. and start BLE library.
  ble_begin();
}

static byte buf_len = 0;

void ble_write_string(byte *bytes, uint8_t len)
{
  if (buf_len + len > 20)
  {
    for (int j = 0; j < 15000; j++)
      ble_do_events();

      buf_len = 0;
```

```
    }

    for (int j = 0; j < len; j++)
    {
      ble_write(bytes[j]);
      buf_len++;
    }

    if (buf_len == 20)
    {
      for (int j = 0; j < 15000; j++)
        ble_do_events();

        buf_len = 0;
    }
  }
}

byte reportDigitalInput()
{
  if (!ble_connected())
    return 0;

  static byte pin = 0;
  byte report = 0;

  if (!IS_PIN_DIGITAL(pin))
  {
    pin++;
    if (pin >= TOTAL_PINS)
      pin = 0;
    return 0;
  }

  if (pin_mode[pin] == INPUT)
  {
    byte current_state = digitalRead(pin);

    if (pin_state[pin] != current_state)
    {
      pin_state[pin] = current_state;
      byte buf[] = {'G', pin, INPUT, current_state};
```

```
      ble_write_string(buf, 4);

      report = 1;
    }
  }

  pin++;
  if (pin >= TOTAL_PINS)
    pin = 0;

  return report;
}

void reportPinCapability(byte pin)
{
  byte buf[] = {'P', pin, 0x00};
  byte pin_cap = 0;

  if (IS_PIN_DIGITAL(pin))
    pin_cap |= PIN_CAPABILITY_DIGITAL;

  if (IS_PIN_ANALOG(pin))
    pin_cap |= PIN_CAPABILITY_ANALOG;

  if (IS_PIN_PWM(pin))
    pin_cap |= PIN_CAPABILITY_PWM;

  if (IS_PIN_SERVO(pin))
    pin_cap |= PIN_CAPABILITY_SERVO;

  buf[2] = pin_cap;
  ble_write_string(buf, 3);
}

void reportPinServoData(byte pin)
{
//  if (IS_PIN_SERVO(pin))
//    servos[PIN_TO_SERVO(pin)].write(value);
//  pin_servo[pin] = value;

  byte value = pin_servo[pin];
  byte mode = pin_mode[pin];
```

```
    byte buf[] = {'G', pin, mode, value};
    ble_write_string(buf, 4);
}

byte reportPinAnalogData()
{
  if (!ble_connected())
    return 0;

  static byte pin = 0;
  byte report = 0;

  if (!IS_PIN_DIGITAL(pin))
  {
    pin++;
    if (pin >= TOTAL_PINS)
      pin = 0;
    return 0;
  }

  if (pin_mode[pin] == ANALOG)
  {
    uint16_t value = analogRead(pin);
    byte value_lo = value;
    byte value_hi = value>>8;

    byte mode = pin_mode[pin];
    mode = (value_hi << 4) | mode;

    byte buf[] = {'G', pin, mode, value_lo};
    ble_write_string(buf, 4);
  }

  pin++;
  if (pin >= TOTAL_PINS)
    pin = 0;

  return report;
}

void reportPinDigitalData(byte pin)
{
  byte state = digitalRead(pin);
```

```
    byte mode = pin_mode[pin];
    byte buf[] = {'G', pin, mode, state};
    ble_write_string(buf, 4);
}

void reportPinPWMData(byte pin)
{
    byte value = pin_pwm[pin];
    byte mode = pin_mode[pin];
    byte buf[] = {'G', pin, mode, value};
    ble_write_string(buf, 4);
}

void sendCustomData(uint8_t *buf, uint8_t len)
{
    uint8_t data[20] = "Z";
    memcpy(&data[1], buf, len);
    ble_write_string(data, len+1);
}

byte queryDone = false;

void loop()
{
    while(ble_available())
    {
        byte cmd;
        cmd = ble_read();
        Serial.write(cmd);

        // Parse data here
        switch (cmd)
        {
            case 'V': // query protocol version
            {
                byte buf[] = {'V', 0x00, 0x00, 0x01};
                ble_write_string(buf, 4);
            }
            break;

            case 'C': // query board total pin count
            {
                byte buf[2];
```

```
      buf[0] = 'C';
      buf[1] = TOTAL_PINS;
      ble_write_string(buf, 2);
    }
    break;

    case 'M': // query pin mode
    {
      byte pin = ble_read();
      byte buf[] = {'M', pin, pin_mode[pin]}; // report pin mode
      ble_write_string(buf, 3);
    }
    break;

    case 'S': // set pin mode
    {
      byte pin = ble_read();
      byte mode = ble_read();

      if (IS_PIN_SERVO(pin) && mode != SERVO &&
      servos[PIN_TO_SERVO(pin)].attached())
        servos[PIN_TO_SERVO(pin)].detach();

      /* ToDo: check the mode is in its capability or not */
      /* assume always ok */
      if (mode != pin_mode[pin])
      {
        pinMode(pin, mode);
        pin_mode[pin] = mode;

        if (mode == OUTPUT)
        {
          digitalWrite(pin, LOW);
          pin_state[pin] = LOW;
        }
        else if (mode == INPUT)
        {
          digitalWrite(pin, HIGH);
          pin_state[pin] = HIGH;
        }
        else if (mode == ANALOG)
        {
          if (IS_PIN_ANALOG(pin)) {
```

```
            if (IS_PIN_DIGITAL(pin)) {
                pinMode(PIN_TO_DIGITAL(pin), LOW);
            }
        }
    }
    else if (mode == PWM)
    {
      if (IS_PIN_PWM(pin))
      {
        pinMode(PIN_TO_PWM(pin), OUTPUT);
        analogWrite(PIN_TO_PWM(pin), 0);
        pin_pwm[pin] = 0;
        pin_mode[pin] = PWM;
      }
    }
    else if (mode == SERVO)
    {
      if (IS_PIN_SERVO(pin))
      {
        pin_servo[pin] = 0;
        pin_mode[pin] = SERVO;
        if (!servos[PIN_TO_SERVO(pin)].attached())
          servos[PIN_TO_SERVO(pin)].attach(
          PIN_TO_DIGITAL(pin));
      }
    }
  }

//        if (mode == ANALOG)
//          reportPinAnalogData(pin);
    if ( (mode == INPUT) || (mode == OUTPUT) )
      reportPinDigitalData(pin);
    else if (mode == PWM)
      reportPinPWMData(pin);
    else if (mode == SERVO)
      reportPinServoData(pin);
  }
  break;

  case 'G': // query pin data
  {
    byte pin = ble_read();
```

```
      reportPinDigitalData(pin);
   }
   break;

   case 'T': // set pin digital state
   {
     byte pin = ble_read();
     byte state = ble_read();

     digitalWrite(pin, state);
     reportPinDigitalData(pin);
   }
   break;

   case 'N': // set PWM
   {
     byte pin = ble_read();
     byte value = ble_read();

     analogWrite(PIN_TO_PWM(pin), value);
     pin_pwm[pin] = value;
     reportPinPWMData(pin);
   }
   break;

   case 'O': // set Servo
   {
     byte pin = ble_read();
     byte value = ble_read();

     if (IS_PIN_SERVO(pin))
       servos[PIN_TO_SERVO(pin)].write(value);
     pin_servo[pin] = value;
     reportPinServoData(pin);
   }
   break;

   case 'A': // query all pin status
   for (int pin = 0; pin < TOTAL_PINS; pin++)
   {
     reportPinCapability(pin);
```

```
    if ( (pin_mode[pin] == INPUT) || (pin_mode[pin] ==
    OUTPUT))
       reportPinDigitalData(pin);
    else if (pin_mode[pin] == PWM)
       reportPinPWMData(pin);
    else if (pin_mode[pin] == SERVO)
       reportPinServoData(pin);
  }

  queryDone = true;
  {
    uint8_t str[] = "ABC";
    sendCustomData(str, 3);
  }

  break;

  case 'P': // query pin capability
  {
    byte pin = ble_read();
    reportPinCapability(pin);
  }
  break;

case 'Z':
  {
    byte len = ble_read();
    byte buf[len];
    for (int i=0;i<len;i++)
      buf[i] = ble_read();
    Serial.println("->");
    Serial.print("Received: ");
    Serial.print(len);
    Serial.println(" byte(s)");
    Serial.print(" Hex: ");
    for (int i=0;i<len;i++)
      Serial.print(buf[i], HEX);
    Serial.println();
  }
}

// send out any outstanding data
```

```
    ble_do_events();
    buf_len = 0;

    return; // only do this task in this loop
}

// process text data
if (Serial.available())
{
    byte d = 'Z';
    ble_write(d);

    delay(5);
    while(Serial.available())
    {
        d = Serial.read();
        ble_write(d);
    }

    ble_do_events();
    buf_len = 0;

    return;
}

// No input data, no commands, process analog data
if (!ble_connected())
    queryDone = false; // reset query state

if (queryDone) // only report data after the query state
{
    byte input_data_pending = reportDigitalInput();
    if (input_data_pending)
    {
        ble_do_events();
        buf_len = 0;

        return; // only do this task in this loop
    }

    reportPinAnalogData();

    ble_do_events();
```

```
    buf_len = 0;

    return;
  }

  ble_do_events();
  buf_len = 0;
}
```

Included with the BLEControllerSketch code is the `<Boards.h>` library. This library provides the definitions and configuration data for the Firmata code. It is an open source code that allows the RedBearLab BLE shield to connect with servo motors, analog sensors, and digital circuits. With the BLEControllerSketch code uploaded to the Arduino, all that remains for the Simple Chat device is to install the BLE Controller Android mobile app. The mobile app can be downloaded from the Google Play store. The following is the BLE Controller app:

 The Firmata software allows the Arduino to control servo motors, digital circuits, and monitor analog sensors using a special **Graphical User Interface (GUI)** panel.

Connecting with an Android smartphone

Follow these steps for connecting the BLE Controller with an Android smartphone:

1. Pressing the **scan** button on the BLE Controller app will start the process of finding and connecting with a Bluetooth-enabled electronic device. To make the process of locating the Android smartphone easy, the RedBearLab BLE shield needs to be paired with the phone. Pairing Bluetooth devices is accomplished under the Android phone's **Setting** panel. Once the pairing process is complete, the BLE Controller app will display the following message on the smartphone's screen:

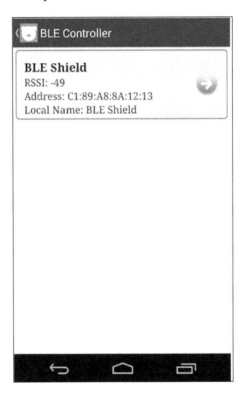

2. Next, open the Arduino IDE serial monitor and set the baud rate for 57,600 bps. The BLE Controller app has several controls and monitor functions to explore, with one of them being the Simple Chat application. Touch the top- left arrow next to the BLE Controller text to obtain the Simple Chat application. The selection application list is shown here:

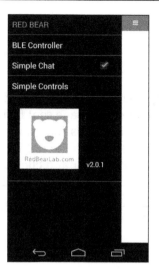

3. Touch the box to open the Simple Chat application. Then, touch the blue arrow in the advertisement window and start typing the message **hi arduino**, as shown here:

When sending a text message, use lower case letters. The code is case sensitive.

4. Touch the send message button and the text will be displayed on the serial monitor. The transmitted text message sent from the Android smartphone to the Arduino IDE serial monitor is shown next. To send messages in the other direction (from the desktop PC or notebook computer), type the text into the Arduino IDE serial monitor and click on **Send**. The text message will now appear on the Simple Chat screen of the Android smartphone.

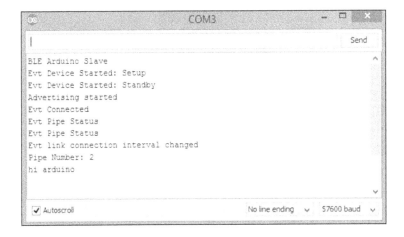

Congratulations, you just built a BLE-based Simple Chat device!

Summary

In this chapter, two Simple Chat device prototypes were built using an Arduino, a 16x2 LCD, a RedBearLab BLE shield, and a littleBits number module. A discussion on hardwired and wireless communications was provided in this chapter. The key technical objectives for this chapter are as follows:

- A prototype serial-based Simple Chat device software and hardware building instructions were presented.

- The 16x2 LCD will display the text message sent by the serial monitor.

- The littleBits electronic number module was introduced in this section as a rapid prototyping tool for serial transmission acknowledgement (validation).

- The Nordic nRF8001 BLE IC internal architecture was discussed using a block diagram. Hardware and software building instructions for a prototype BLE based Simple Chat device were provided within the second project.

- The RedBearLab BLE shield's basic functions were discussed using a block diagram. Instructions for adding the RBL_nRF8001 library to the Arduino IDE was explained.

- The RedBearLab URL website was provided as an additional resource for properly installing the RBL_nRF8001 library on to the Arduino IDE.

- The BLE Controller app for Android was downloaded from the Google Play store and installed on an Android smartphone.

- The final test of the project consisted of typing a message onto the Simple Chat device and sending it to Android smartphone.

- Wiring diagrams and circuit schematic diagrams for the electronics for both Simple Chat devices were provided and discussed in this chapter.

In the next chapter, the RedBearLab BLE shield will be discussed some more, but this time as a BLE controller. A discussion on how to control a 7 segment LED display using the Arduino Uno's digital ports with a BLE Controller mobile app will be provided. Learning objectives in the next chapter will include the BLE Controller setup, wiring the 7 segment LED display to the Arduino, and creating numbers and letters on the 7 segment LED display using the BLE controller.

As usual, mini lab testing procedures, Arduino code explanation, and the final project assembly will be discussed in the next chapter.

7
Bluetooth Low Energy Controller

Bluetooth technology is one of the most common standards for electronic products communicating within short-range distances. Its data exchange uses a short wavelength, and it falls within the **Instrument Scientific Medical (ISM)** band from 2.4 to 2.485 GHz. It's a wireless technology standard that was invented in 1994 by telecom vendor Ericsson and was designed as an alternative to RS2322 serial cables. Bluetooth technology can connect a multitude of devices and eliminates the problems of communication synchronization exhibited by the RS232 technology.

 RS232 and serial are the technical words used interchangeably in the field of electronic communications.

The **Institute of Electrical and Electronics Engineers** (**IEEE**) standardized Bluetooth as IEEE 802.15.1 but has also released engineering responsibilities to the Bluetooth **Special Interest Group** (**SIG**) organization. The next evolution of the wireless technology is documented in the Bluetooth version 4.22 specifications. The version 4.22 specification now stipulates the integration of a microcontroller to manage the power usage of the wireless device. One of the key elements of the specification is to assure that Bluetooth products being developed will be able to operate on a minimum of a 3 V coin cell battery, as shown in the following screenshot, for several months to a year. Therefore, the new communications standard is titled Bluetooth Low Energy Smart.

 For additional information on Bluetooth technologies, visit the website at `https://developer.bluetooth.org/TechnologyOverview/Pages/BLE.aspx`.

In this chapter, we will learn how to build a **Bluetooth Low Energy** (**BLE**) wireless controller. As in *Chapter 6, A Simple Chat Device with LCD*, the Arduino Uno will serve as the main processing device. The littleBits motor module and a seven segment LED display will be the two example devices to control LED by BLE. The RedBearLab BLE shield and the BLE Controller mobile app will provide visual and wireless communication features to enhance the user interface experience. In addition, an Android smartphone and the RedBearLab BLE mobile app (BLE Controller) will be used to send control messages to the BLE shield. If additional information is needed on the operation of the Nordic nRF8001 BLE IC, please refer to the previous chapter. The parts required to build the BLE Controller are listed as follows.

Parts list

Following is the list of parts required for the BLE Controller:

- Arduino Uno (one unit)
- littleBits motor module (one unit)
- littleBits proto board (one unit)
- Seven segment LED display (Common Anode) (one unit)
- RedBearLab BLE shield (one unit)
- 330 ohm resistor (orange, orange, brown, and gold stripes) (one unit)
- Breadboard
- Wires
- Android smartphone

BLE Controller block diagrams

The BLE Controller is a wireless device, capable of easily turning *ON* and *OFF* electronic circuits and electromechanical components. A wireless packet of digital commands are sent from an Android smartphone and received by the BLE Controller. The RedBearLab BLE Controller app is used as the **User Interface** (**UI**) to select the digital ports within the connected electronic circuits or electromechanical components wired to the Arduino Uno. Based on the connected electronic circuit or electromechanical component needing a logic **HIGH** or **LOW** control signal to operate, the selected digital port can be configured quite easily using slide switches. The concept drawing of the BLE Controller is shown in the following diagram:

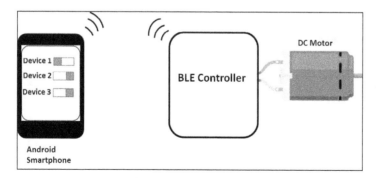

The BLE Controller block diagram is an engineering development tool used to convey complete product designs using graphics. The block diagram also makes it easier to plan the breadboard for prototyping and testing of the BLE Controller in a Maker's workshop or laboratory bench. The final observation of the BLE Controller block diagram is the basic computer convention of inputs on the left-hand side, the processor located in the middle, and the outputs placed on the right-hand side of the design layout. As shown, the Android smartphone is on the left-hand side, the Arduino with RedBearLab shield is located in the middle, and the seven segment LED display and the littleBits motor module are shown on the right-hand side of the block diagram. The littleBits motor module will turn *ON* and *OFF*, based on receiving the wireless packet of digital commands from the BLE Controller mobile app. Selecting the slide switches in a digital pattern will display numbers and letters on the seven segment LED display. Following is the block diagram of BLE Controller with littleBits motor module:

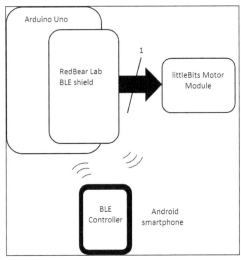

BLE Controller wired to a littleBits motor module block diagram

 Wireless is the marketing word for **Radio Frequency (RF)**.

The block diagram of BLE Controller with seven segment LED display is as follows:

BLE Controller wired to a seven segment LED display block diagram

Building a BLE DC motor controller

The BLE Controller can easily be hardwired to a DC motor. The littleBits DC motor module is a combination driver circuit that provides signal buffering and bidirectional control by way of a small selector switch. The small electronics module provides convenience and ease of wiring the DC motor to the RedBearLab shield. The actual DC motor driver circuit component used to operate the electrically connected small electromechanical device is a power MOSFET (TPS2812C).

The TPS2812C is an 8-pin **small outline integrated circuit** (**SOIC**) using dual high-speed power MOSFET drivers capable of delivering 2 A (amperes) of peak current to an electrical load. As a reference, the circuit schematic diagram of the littleBits DC motor is shown as follows:

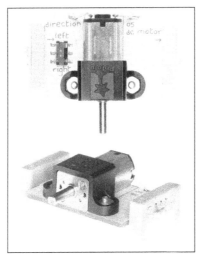

The littleBits DC motor module

 BLE is also known as **Bluetooth Smart**.

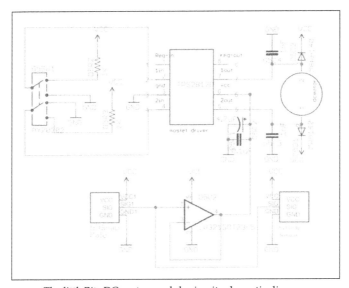

The littleBits DC motor module circuit schematic diagram

A MOSFET has an insulating layer of oxide substrate material to prevent transistor operation instability from the leakage current, **electrostatic discharge** (**ESD**), or unwanted electrical noise (voltage transients). The transistor's gate pin is insulated from the channel by this oxide substrate material. Electrons fill up the channel based on the voltage being applied to the MOSFET's gate pin. The internal capacitance of the gate pin allows the transfer of these electrons into the channel. The amount of electrons in the channel will turn on the MOSFET, which allows it to operate a DC motor, an electromechanical relay, or an LED wired to its drain pin.

The BLE Controller is a two-part wireless communication device that consists of a RedBearLab BLE shield and the Arduino Uno. The RedBearLab BLE shield sits on top of the Arduino Uno, making a compact wireless electronic controller. Single inline female header connectors are used to make wiring connections to the littleBits DC motor module using jumper wires. A littleBits proto module provides an electrical wiring interface to connect the RedBearLab shield to the DC motor. An electrical wiring diagram of the BLE Controller wired to a littleBits DC motor is shown as follows:

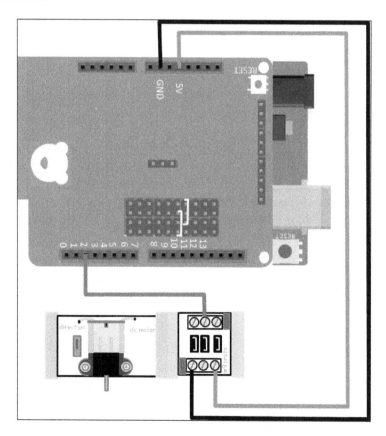

As an additional wiring reference aid, the actual prototype BLE motor controller is shown as follows:

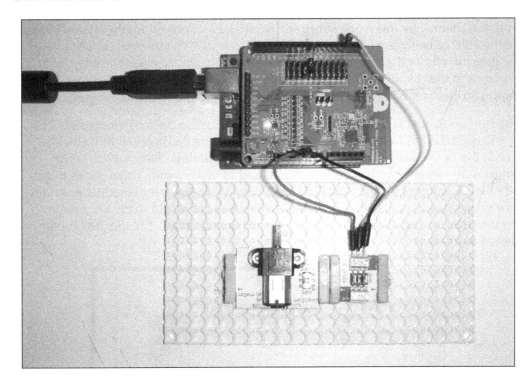

The Arduino code discussed in *Chapter 6, Simple Chat Device with LCD* in section *Uploading the BLEControllerSketch code to the Arduino Uno* will be programmed in the BLE Controller as well. After the code is uploaded to the Arduino Uno, the BLE Controller mobile app discussed in the previous chapter will operate the littleBits DC motor module. Go to the screen of the BLE Controller, which shows the majority of pins on the Arduino Uno later.

 The littleBits DC motor could be replaced by other littleBits output modules instead.

Advancing the smartphone screens for setup and control of a littleBits DC motor is as follows:

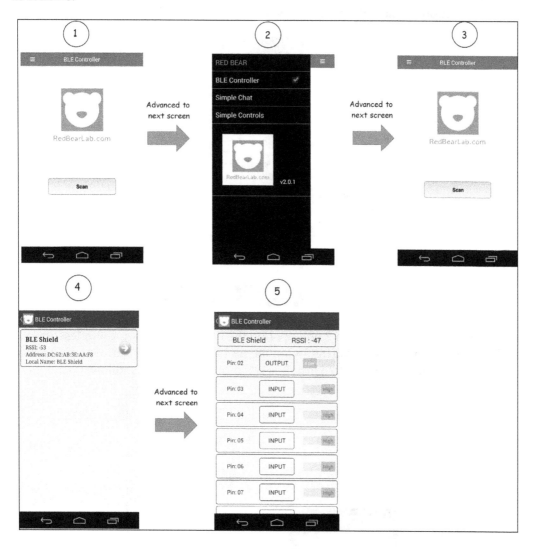

Pin 2 is changed to an output pin by touching the **INPUT** button on the smartphone's touch screen. A small window will appear at the bottom of the screen. Touch the **OUTPUT** button to make the change as shown in the following diagram:

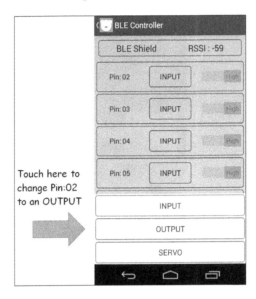

Sliding the digital switch of Pin 2 will operate the littleBits DC motor wirelessly via the BLE Controller. The diagram showing the mobile app on an Android smartphone and the BLE Controller wired to a littleBits DC motor is shown as follows:

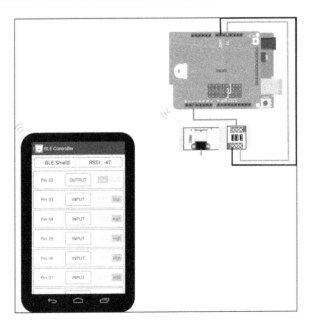

Congratulations, you have just built a BLE-Controlled DC motor! Now, the information presented for this project will be used to build a wireless controller to operate a seven segment LED display.

 If the slide switch and littleBits DC motor control function stop working (times out), close the BLE Controller app. Reopen the application and toggle the slide switch to operate the littleBits DC motor again.

Building a BLE seven segment LED display controller

The prototype built for this wireless device is quite similar to the BLE DC motor controller. The smartphone app will be extended in such a way that six of the digital ports will be used instead of one. Traditionally, to operate seven discrete LEDs of the display requires the same number of control lines to operate the optoelectronic component properly. The BLE Controller app unfortunately has only six slide switches; therefore, the extra LED of the display will be hardwired to a slide switch-controlled segment. The BLE seven segment LED display prototype concept's diagram is as follows:

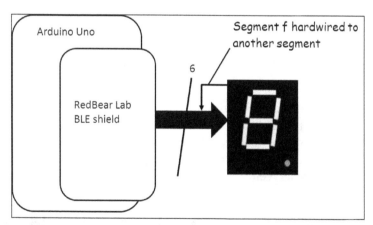

What's inside a seven segments LED display?

Before proceeding to the project build, let's explore the inner workings of a seven segment LED display. A seven segment LED display consists of seven discrete LEDs strategically arranged inside a small plastic case. The placement of the discrete LEDs is such that the number 8 is visible on the front of the small plastic case.

Also, the arrangement of the discrete seven LEDs is such that two types of displays are made: **Common Anode** and **Common Cathode**. The common anode seven segment LED display is the one in which all of the discrete LEDs' anodes are connected together. The common anode pin is connected to a +Vcc DC power supply. To prevent the discrete LEDs from getting damaged, a current limiting resistor is wired between the +Vcc DC power supply and the common anode pin of the seven segment LED display. An example of a common anode seven segment LED display is shown here:

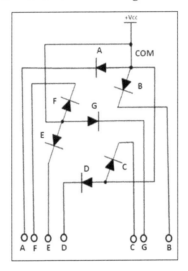

All of the seven segment LED cathodes are wired to ground. The discrete LEDs are then turned *ON*, displaying the number 8. Some letters (A, C, E, F, H, J, L, P, U, h) and whole numbers (0-9) can be easily displayed by wiring the appropriate LED segments to ground.

In addition, common cathode is the other type of seven segment LED display. All of the cathodes are wired together and connected to ground. The anodes of the discrete LED segments are wired to a positive voltage source through series limiting resistors. The letters and numbers for the common anode LED display can be wired on a common cathode device as well.

Wiring the seven segment LED display to the BLE shield

The final build of the BLE-Controlled seven segment LED display consists of wiring the RedBearLab BLE shield to the optoelectronic component. A tool that makes it easy to wire seven segment LED displays to electronic circuits (analog, digital, and microcontroller) is a pinout sheet.

The pinout sheet is a table or component diagram that shows the pin number, pin description, and electrical connection for each LED segment. An example of a pinout sheet for a common anode seven segment LED display is shown in the following screenshot:

Pin no.	Electrical connection
1	Cathode A
2	Cathode F
3	Common Anode
4	No Pin
5	No Pin
6	Cathode D. P.
7	Cathode E
8	Cathode D
9	No Connection
10	Cathode C
11	Cathode G
12	No Pin
13	Cathode B
14	Common Anode

This pinout sheet will be used in wiring the seven segment LED display to the RedBearLab BLE shield. Refer to this page as a reference tool when wiring the RedBearLab BLE shield to the common anode seven segment LED display. To aid in the placement of the seven segment LED display and 330 ohm resistor onto the breadboard, an electrical wiring diagram is shown as follows:

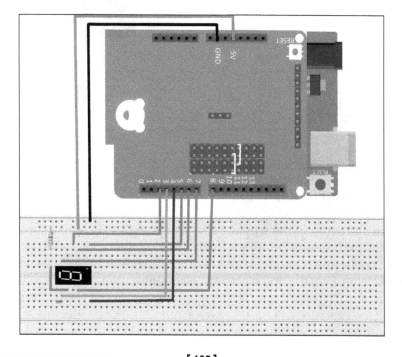

MAN72 is the name used to reference a common anode seven segment LED display. The MAN74 is the name used for a common cathode seven segment LED display.

A common cathode seven segment LED display can easily be used in this project by connecting pin 3 to ground. Six 330 ohm series limiting resistors will be wired between the Arduino Uno designated digital pins and seven segment LED anodes. As a reference drawing to aid in the wiring of the common anode seven segment LED display to the RedBearLab shield, a circuit schematic diagram is provided as follows:

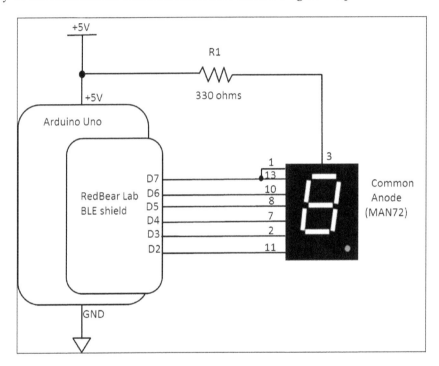

 A 16-pin **Dual inline package** (**DIP**) 330 ohm resistor pack can replace the individual resistors to aid in the wiring of a common cathode seven segment LED display project.

The operation and setup of the wireless seven segment LED display controller is the same as for the littleBits DC motor device. For reference, the setup of the smartphone mobile app is shown later in the screenshot. Set the slide switches to display the letter *F* on the optoelectronic electronic component.

As discussed earlier, since the RedBearLab BLE shield is able to accommodate six discrete LEDs of the seven segment LED display, the seventh pin (cathode−*f*: [pin 1]) is wired to pin 13 of the MAN 72 component. The final prototype BLE seven segment LED display controller is provided.

To test the seven segment LED display wiring prior to the mobile app control, touch each pin to ground. If each LED segment turns on, the wiring is correct.

Reference diagram for advancing the smartphone screens for setup and control of a littleBits DC motor is shown here:

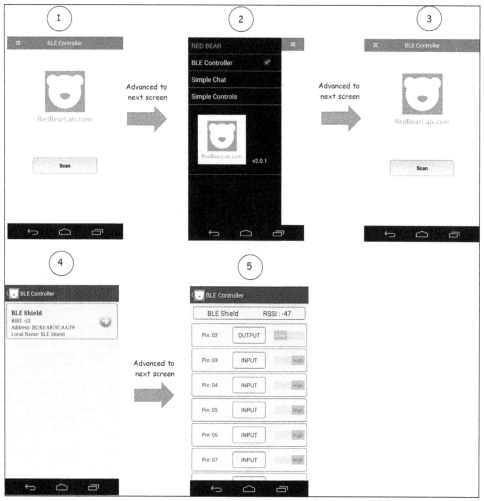

Reference diagram: Advancing the smartphone screens for setup and control of a littleBits DC motor.

Reference diagram for changing an INPUT pin to an OUTPUT pin is as follows:

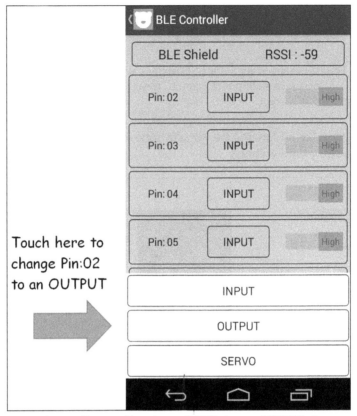

Reference diagram: Changing an INPUT pin to an OUTPUT pin.

Try creating a new number or a letter by setting the appropriate slide switches of the BLE Controller mobile app to operate the right LED segments. The example is shown in the following diagram:

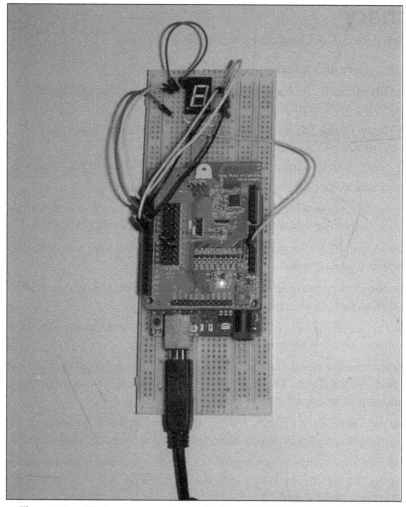

The prototype BLE seven segment LED display controller displaying the letter *F*.

Congratulations on successfully building a BLE-Controlled seven segment LED display controller!

Summary

In this chapter, two BLE Controller prototypes were built:

- A DC motor and a seven segment LED display.

- The littleBits DC motor electronic control circuit was discussed by way of reviewing the circuit schematic diagram.

- The heart of the DC motor electronic control circuit is a power MOSFET (part number TPS2812C).

- The TPS2812C is a dual high-speed power MOSFET driver capable of providing peak currents of 2 A to an electrical load.

- The smartphone screen setup and control to operate the littleBits DC motor with the mobile app was discussed and provided as well.

- The second BLE Controller prototype to operate a seven segment LED display was discussed.

- The internal construction of a seven segment LED display was presented in this section of the chapter.

- Two types of the seven segment LED displays — common cathode and common anode — were explained.

- The common cathode seven segment LED display has all cathodes tied to one electrical pin. The common anode component has all of the anodes tied to one electrical pin.

- The pinout sheet is a table or diagram showing the pin numbers, pin descriptions, and electrical connections of a seven segment LED display.

- The common anode seven segment LED display was configured to display the letter *F* by way of the setting slide switches on the BLE Controller mobile app.

In *Chapter 8, Capacitive Touch Sensing* will be discussed. A capacitive touch sensor using a 555 timer and an Arduino Uno will be presented in the chapter. Although the 555 timer is a legacy IC, it has a wealth of applications in electronic products. Learning objectives in the chapter are as follows:

- The 555 timer one-shot function

- How to make a touch plate with common household items

- How to wire the touch plate to the 555 one-shot timer

- How to wire the one-shot timer to the Arduino Uno
- How to wire the small DC motor driver circuit using discrete electronic components and a littleBits DC motor module to the Arduino Uno.

As usual, mini lab-testing procedures, Arduino code explanations, and the final project assembly will be discussed in next chapter.

8

Capacitive Touch Sensing

In today's industrial environments, manufacturing plants are using touch screen panels to operate robotic work cells. By touching graphic buttons on a flat panel display, an operator can easily control the movement of a robot. The interaction between the operator's hand and touch screen creates a tiny capacitor. The microcontroller reads the capacitance value and allows the programmable chip to control a robot or industrial machine. Another sensing technology used in touch screens is **resistive touch**. Resistive touch technology is based on two flexible resistive sheets separated by air. Touching the screen allows a voltage drop to be produced by the flexible resistive sheets. The voltage value is mapped to the screen's touch location. Today, some microcontrollers include capacitive touch sensing technologies within their silicon substrate. Although the Arduino doesn't have a capacitive sense feature embedded within its silicon substrate, it can easily be added externally using a fewoff-the-shelf electronic components.

An example of a consumer product that uses capacitive touch sensing is a **pen stylus**. A standard pen stylus can be made into a capacitive touch sensing device by adding a conductive surface. When a part of our body (finger or hand) makes contact with the conductive surface, a small capacitor element is formed. The microcontroller can read this capacitive value allowing the instrument to engage with the images on a touch screen. Examples of capacitive pen stylus are shown in the following image:

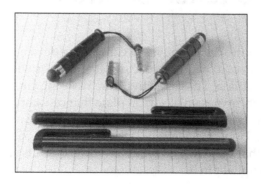

In this chapter, you will learn how to make a capacitive sensor to control a servo motors angular rotation by building a **capacitive touch controller**. A 555 timer IC along with a capacitor detection circuit will provide the trigger control for the Arduino to drive the servo motor's direction. The parts required to build the capacitive touch controller are listed as follows:

Parts list

The following is the list of parts required to build a apacitive Touch controller:

- R1 – 1 MΩ resistor (brown, black, green, and gold stripes)
- R2 – 1 KΩ resistor (brown, black, red, and gold stripes)
- R3 – 100 KΩ resistor (brown, black, yellow, and gold stripes)
- C1 – 10 µF electrolytic capacitor
- C2 – 0.001 µF or 1 nF capacitor
- Q1 – 2N3904 NPN transistor
- DC servo motor (3-6 V rated) (one unit)
- Arduino Uno (R3 version) or equivalent
- littleBits DC motor
- littleBits proto module
- Aluminum foil
- Cardboard
- Tape
- A sheet of clear plastic
- Electrical wire (2" in length with a 1/4" of insulation removed from both ends) (one unit)

A capacitive touch controller block diagram

The capacitive touch controller is a physical computing device capable of controlling a servo motor's angular rotation. The capacitive interface provides an *ON* and *OFF* control signal to the Arduino. The capacitive touch sensing circuit works with human body capacitance which creates an electrical plate that can store or release static energy. The change in static energy within the electrical plate is used to trigger the 555 timer detection circuit. Upon receiving the static energy, the 555 timer IC generates a small electrical pulse that the Arduino reads and then outputs an angle control signal to the servo motor at digital pin 9.

The servo motor's direction changes when touching the capacitive touch sensor electrode (wire). To make the capacitive touch sensor easy to use, the electrode is attached to a metal plate for touch control. A concept drawing of the capacitive touch controller with the sensor is as follows:

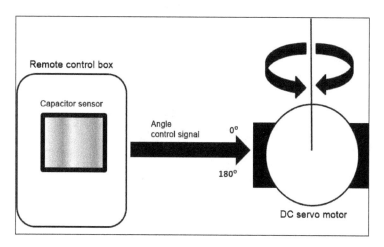

The capacitive touch controller block diagram is an engineering development tool that is used to convey complete product designs using graphics. The block diagram also makes it easy to plan the breadboard for prototyping and testing of the capacitive touch controller in a maker's workshop or laboratory bench. The final observation of the capacitive touch controller's block diagram is the basic computer convention of the inputs on the left-hand side, the processor located in the middle, and the outputs placed on the right-hand side of the design layout. As shown in the figure, the capacitor-touch sensor circuit is on the left-hand side, the Arduino is located in the middle, and the DC servo motor is shown at the right of the block diagram. The servo motor will change direction (forwards or backwards) based on the capacitor sensor detecting a touch. The servo motor will return to the forward direction after the touch has been released:

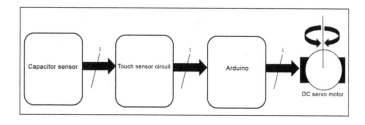

 The 555 timer IC is an 8-pin **Integrated Circuit** (**IC**) used in a variety of timer, switching, pulse generation, and oscillator applications.

Building a capacitive touch controller

The capacitive touch controller can easily be hardwired to a DC servo motor. The touch sensor circuit provides the electronics for reading the capacitances of the sensing plate and provides an Arduino control signal to operate the DC servo motor. The steps to build the capacitive touch controller are as follows:

1. Build a capacitor sensor using the assembly diagrams as follows:

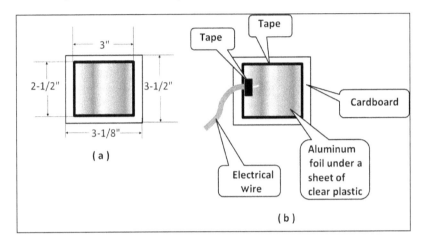

[
 Image (a) shows the capacitor sensor dimensions, while Image (b) shows the individual materials assembled.
]

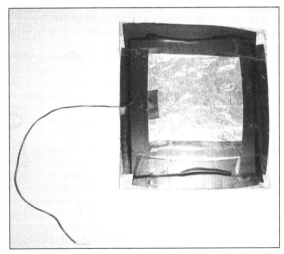

The finished capacitor touch plate sensor

Wire the capacitive touch controller circuit onto a solderless breadboard as shown in the following diagram. The circuit schematic diagram is provided as an additional wiring aid to build the capacitive touch controller prototype as follows:

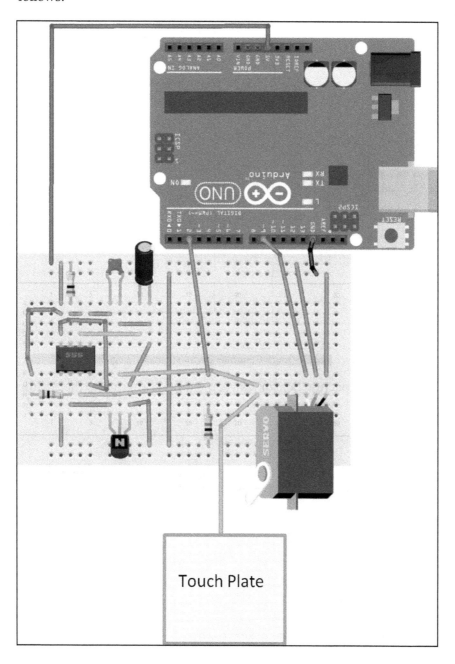

Touch Plate

The colors shown in the wiring diagram are based on the **Fritzing servo motor** module. Obtain the manufacturer's datasheet on the servo motor being used in your capacitive touch controller prototype build. Wire the servo motor to the Touch controller circuit according to the manufacturer's datasheet.

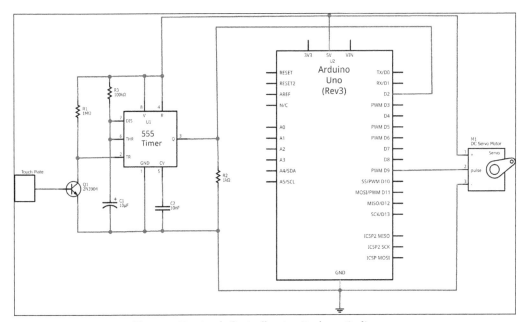

The capacitive touch Controller circuit schematic diagram

2. Upload the Capacitor_Sensor_2sweep_motion program to the Arduino with the sketch as follows:

```
/*
  Capacitor_Sensor_2sweep_motion

Sweeps a servo motor counter-clockwise and clockwise based
on touching a capacitor sensor (touch plate).

07 July 2013
by Don Wilcher

*/

#include <Servo.h>

Servo myservo;  // create servo object to control a servo
```

```
// constants won't change. They're used here to set pin
Arduino pin numbers

const int sensorPin = 2;      // the number of the sensor
pin
int sensorStatus = 0;    // variable to store the servo
position
int pos = 0; // variable to store servo motor angles

void setup() {
  // initialize the sensor pin as an input:
  pinMode(sensorPin, INPUT);

  myservo.attach(9);   // attaches the servo on pin 9 to the
  servo object
}

void loop(){
  // read the status of the sensor value:
  sensorStatus = digitalRead(sensorPin);

  // check if the sensor is activated.
  // if it is, the sensorStatus is HIGH:
  if (sensorStatus == HIGH) {
    // rotate servo motor in CW direction:
    for(pos = 0; pos < 170; pos += 1)   // goes from 0
    degrees to 180 degrees
    {                                   // in steps of 1
    degree
    myservo.write(pos);                 // tell servo to go to
    position in variable 'pos'
    delay(15);
    }
  }
  else {
    // rotate servo motor in CCW direction:
    for(pos = 170; pos>=1; pos-=1)     // goes from 180
    degrees to 0 degrees
    {
    myservo.write(pos);                 // tell servo to go to
    position in variable 'pos'
    delay(15);                          // waits 15ms for the
    servo to reach the position
    }
  }}
```

3. After uploading the sketch to the Arduino, the servo motor should begin sweeping in a clockwise rotation.

4. Momentarily touch the capacitor sensor (touch plate) with your hand until the servo motor sweeps in a counterclockwise rotation.

5. Continue touching the capacitor sensor and watch the servo motor sweep in both the directions. The time delay for sweeping the servo motor's directions is 15 **milliseconds** (**ms**).

Congratulations on successfully building a Capacitor Touch controller!

The 555 timer IC's monostable operation

The main electronic component for the capacitive touch sensor circuit is the **555 timer IC**. As shown in the circuit schematic diagram, the 555 timer is wired as a triggered monostable timer circuit. The monostable circuit is quite simple in construction, and it consists of three timing electronic components — the 555 timer IC, R3, and C1. The bypass capacitor (C2) helps minimize noise from the timer's internal switching circuits. The pull-down resistor (R2) is included to assure that 0 V DC is applied to the Arduino pin D2 when a trigger signal is not present. The transistor (Q1) and the pull-up resistor (R1) provide an inverting electronic switch, which is used to trigger the 555 timer monostable circuit. The 555 timer-triggered monostable circuit is shown in the following diagram:

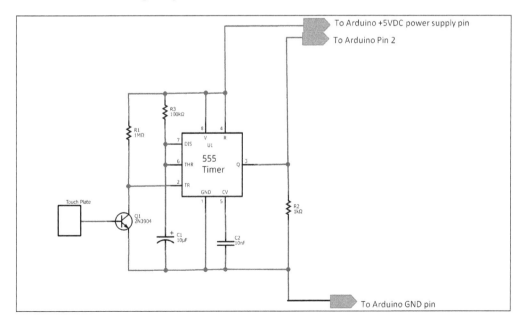

Another term used to describe a monostable circuit is one-shot timer.

To trigger the 555 timer monostable circuit, either zero or negative voltage is required at pin 2 of the timing IC. The transistor switch provides a 0V DC signal upon touching the capacitor touch plate. When the 0V DC signal appears at the 555 timer trigger input, the output at pin 3 of the IC rises to a high voltage approaching +5V DC power supply. A triggered 555 timer monostable's output response is shown in the simulation model given later. The 7404 Hex Inverter IC provides the 0V DC signal required to trigger the 555 timer monostable circuit. Therefore, this output response shown in the simulation model represents a hand touching the capacitor touch plate.

 As shown in the circuit schematic diagram, Touch-Plate is another name for capacitor sensor.

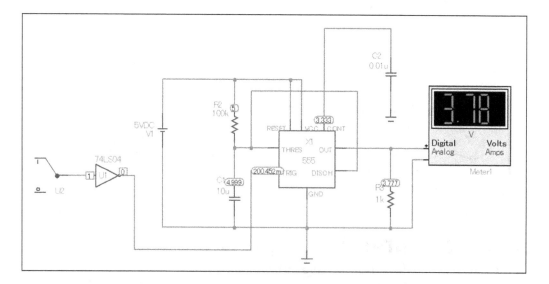

In addition, the C1 (10 μF) electrolytic capacitor begins to charge toward the +5V DC power supply. When the voltage across C1 reaches two-thirds of the +5V DC power supply, the electrolytic capacitor discharges after completing the timing cycle. The output at pin 3 is approximately 0 V DC. The results of this simulation are shown in the following diagram. The output response shown in the simulation model represents no hand detection from the capacitor touch plate:

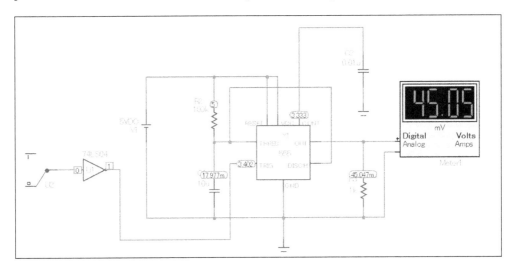

The simulation models were built using Micro-Cap circuit simulation software. A free download of the software can be found at the URL address of `http://www.spectrum-soft.com/download.shtm`. A subscription-based circuit simulation package for Mac OS X and Linux machines can be obtained at the URL address of `https://www.circuitlab.com/`.

A Do It Yourself Design Challenge!

Industrial control panels usually have indicators to provide visual feedback on key electromechanics and circuits of a robotic system. An LED can be added to the Capacitor Touch controller to provide hand detection plate visual feedback. The LED turns on when a hand touches the sensor and turns off when it is removed. You can use the onboard LED wired to pin D13 of the Arduino or add an external component. Here are the lines of code to accomplish the task:

```
int sensorStatus = 0;     // variable to store the servo
position
int ledPin = 13; // the number of the LED pin (add this line of
code)
```

```
void setup() {
  // initialize the sensor pin as an input:
  pinMode(sensorPin, INPUT);
  pinMode (ledPin, OUTPUT); (add this line of code)

if (sensorStatus == HIGH) {
  // turn ON LED
  digitalWrite (ledPin, HIGH); (add this line of code)

}
```

The modified Arduino code for the LED indicator function is shown later. Save the code changes with a different name to reflect the software revision. Upload the new code to the Arduino and notice the LED turning on when momentarily touching the capacitor sensor and turning off a few seconds later. To include an electromechanical actuator in the design challenge, add the littleBits DC motor. The littleBits DC motor can be controlled by this modification by wiring it to pin 13. The proto module discussed in the previous chapter will aid in wiring the small DC motor to Arduino pin 13. The partial wiring diagram showing the littleBits DC motor and proto module connected to an Arduino is shown in the following image. Record the new code, littleBits DC motor, and LED circuit in a lab notebook:

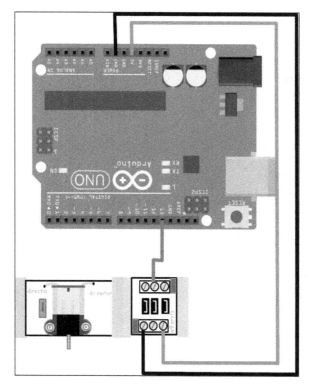

The code for Capacitor Touch controller with LED is as follows:

```
/*
Capacitor_Sensor_2sweep_motion with LED indicator

Sweeps a servo motor counter-clockwise and clockwise based on
touching a capacitor sensor (touch plate).
LED added for capacitor sensor touch feedback


07 July 2013
by Don Wilcher

 */


#include <Servo.h>
Servo myservo;  // create servo object to control a servo

// constants won't change. They're used here to set pin Arduino
pin numbers

const int sensorPin = 2;     // the number of the sensor pin
int sensorStatus = 0;    // variable to store the servo position
int ledPin = 13; // the number of the LED pin
int pos = 0; // variable to store servo motor angles

void setup() {
  // initialize the sensor pin as an input:
  pinMode(sensorPin, INPUT);
  pinMode (ledPin, OUTPUT);

  myservo.attach(9);  // attaches the servo on pin 9 to the servo
  object
}

void loop(){
  // read the status of the sensor value:
  sensorStatus = digitalRead(sensorPin);

  // check if the sensor is activated.
  // if it is, the sensorStatus is HIGH:
  if (sensorStatus == HIGH) {
```

```
// turn ON LED
digitalWrite (ledPin, HIGH);
// rotate servo motor in CW direction:
for(pos = 0; pos < 170; pos += 1)   // goes from 0 degrees to
180 degrees
  {                                 // in steps of 1 degree
  myservo.write(pos);               // tell servo to go to
  position in variable 'pos'
  delay(15);
  }
}
else {
  // turn OFF LED
  digitalWrite (ledPin, LOW);
  // rotate servo motor in CCW direction:
  for(pos = 170; pos>=1; pos-=1)    // goes from 180 degrees to
  0 degrees
  {
    myservo.write(pos);             // tell servo to go to
    position in variable 'pos'
    delay(15);                      // waits 15ms for the servo
    to reach the position
  }
 }
}
```

Before uploading the code to the Arduino, remove the D9 pin wire of the Arduino from the servo motor. The USB port will not see a power surge due to the additional electrical load the servo motor adds to the desktop PC or notebook computer. Reconnect the wire after the upload of the code to the Arduino is complete.

 Changing capacitor C1 to a bigger electrolytic value (that is, 100 µF) will allow a longer sweeping motion delay for the servo motor after removing your hand from the touch plate.

A buttonless servo motor controller

One touch sensor to control two servo motor sweeping motions is fine, but having an additional touch plate provides an efficient human-machine interface to control robots or industrial machines. As a design challenge, create a buttonless servo motor controller where two touch plates (capacitor sensors) can be used independently to sweep the servo motor clockwise or counterclockwise. Build a second capacitor sensor using the block diagram shown here as your reference assembly diagram:

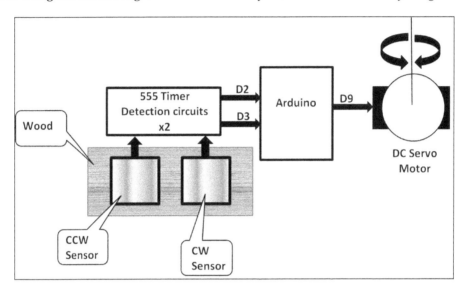

Wire another 555 timer monostable circuit onto the breadboard using the wiring diagram shown in *Building a Capacitive Touch controller* section of this chapter for reference. Also, modify the original code shown earlier where two LEDs provide visual direction of the servo motor when touching either CCW or CW touch sensors.

For convenience, mount each capacitor sensor on an appropriate-sized piece of wood, foam core, or cardboard, as shown in the reference assembly diagram. As mentioned previously, document all wiring diagrams and Arduino code in your lab notebook.

Forrest Mims has written a nice article for **Make Magazine** on how to document your technical and scientific discoveries. Here's the website URL of this article: http://makezine.com/projects/make-34/country-scientist-how-to-document-what-you-make-or-discover/.

Summary

In this chapter, a capacitive touch controller prototype was built:

- A discussion on assembling a capacitor touch plate using household materials was presented

- A procedure to build the capacitive touch controller was provided in this chapter as well

- Uploading the capacitive touch controller code to the Arduino was included

- The explanation of the operation for the code was provided within the comment statements

The electrical operation of a 555 timer-triggered monostable circuit was discussed. To aid in the discussion of the 555 timer-triggered monostable device, Micro-Cap circuit simulation models were provided. A design challenge was presented where the LED mounted on the Arduino board was used as a visual indication of the servo motor's direction. Also, the littleBits DC motor wiring diagram was included to aid in wiring an electromechanical component to the Arduino. Upon hand detection by the capacitor sensor, the littleBits DC motor will turn on. As a final design challenge, a buttonless servo motor control was discussed where two capacitive touch sensors were used to control the direction of a small DC motor. The assembly diagram to build the buttonless controller was provided. Additional information for wiring a second 555 timer along with capacitor sensor construction suggestions was included as well.

In *Chapter 9, Arduino-SNAP Circuit AM Radio* will be discussed. The Elenco SNAP circuit kit uses brightly colored electronic shapes that "snap" together to make all sorts of electronic gadgets. The Arduino can be used as an enable switch for the AMA radio module along with using the RedBearLab shield and Android smartphone. An LCD will be used to display "Radio On / Radio Off messages for the AMA radio operation". Learning objectives in the chapter are to include the following:

- How to make an Arduino to SNAP circuit interface

- The SNAP circuit AMA radio module I/O

- Testing the Arduino-SNAP circuit AMA radio

- How to wire the one-shot timer to the Arduino Uno

As usual, mini lab-testing procedures, Arduino code explanations, and the final project assembly will be discussed in the next chapter.

9
Arduino-SNAP Circuit AM Radio

The **superheterodyne** (**superhet**) radio is considered to be the first receiver using AM for audio reception. Edwin Howard Armstrong, an inventor, is credited with the creation of many of the AM radio features known today. Edwin Armstrong patented three important functions that made the use of AM radio possible today: regeneration, the superhet circuit, and positive feedback. Regeneration allows an electronic signal, such as radio waves, to be amplified many times by the same active device, such as a transistor or operational amplifier. The superhet circuit can easily convert a received signal to a fixed intermediate frequency (IF), which can be more conveniently processed than the original radio carrier frequency. Finally, positive feedback increases the gain or amplification of radio waves by using a closed-loop amplifier circuit. An example of a superhet AM radio is shown in the following screenshot:

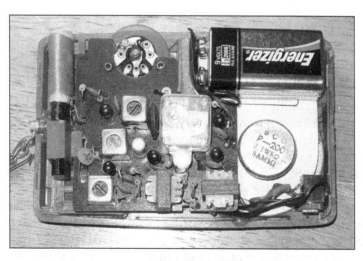

Superhet is short form for superheterodyne. AC is the abbreviation for alternating current.

In this chapter, you will learn how to build an AM radio using the SNAP circuit kit. Also, you will learn how to control the SNAP AM radio using a littleBits IR remote sensor and an ordinary IR handheld remote. An Arduino will be used as the control interface for the SNAP circuit AM radio and LCD. The LCD will display the ON/OFF operation of the SNAP AM radio. A RedBearLab BLE-SNAP circuit AM radio controller will be presented as a DIY challenge as well. The parts required to build the Arduino-SNAP circuit AM radio are listed as follows.

Parts list

The list of parts required to build an Arduino-SNAP circuit AM radio is as follows:

- SNAP Circuits Extreme SC-750 Electronics Discovery Kit (one unit)
- Arduino Uno (R3 version) or equivalent (one unit)
- littleBits remote trigger (one unit)
- littleBits latch (one unit)
- littleBits proto module (one unit)
- littleBits mounting board (one unit)
- R1 – 470 Ω resistor (yellow, violet, brown, and gold stripes)
- D1 – 1N4001 general purpose diode
- K1 – 5V DC electromechanical relay (SPDT type)
- Q1 – 2N3904 NPN transistor
- DIS 16x2 LCD

Radio communication basics

The radio waves are actually high-frequency AC signals. An antenna wired to a high-frequency oscillator can transmit radio waves. The oscillator's frequency is usually 100 KHz or higher. In addition to transmitting high-frequency AC signals, the antenna converts the radio waves into electromagnetic waves as well. Electromagnetic waves travel at the speed of light, which is approximately 3×108 m/s. Upon an antenna receiving a transmitted electromagnetic wave, high-frequency current is created. The created current in the antenna is just a fraction of the transmitting signal. The information in this electromagnetic wave has some electrical signal data, such as frequency, voltage field, magnetic field, and waveform characteristics (sine, square, triangle, and so on). An example of an electromagnetic wave is shown here:

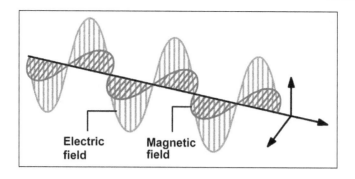

Electric field

Magnetic field

Modulation

In order for audio information (music, speech, and so on) to be transmitted over long distances, an electronic signal mixing technique must be used. The process of superimposing information in an electromagnetic wave is called **modulation**. To illustrate this radio communication technique, take the example of a high-frequency oscillator producing a signal of 500 KHz. This high-frequency signal is called the carrier. The audio oscillator produces an audio signal of 1 KHz. The carrier is modulated with the audio signal in an electronic circuit named a **modulator**. The output of the modulator is connected to the antenna, where the superimposed high-frequency AC signal is converted into a modulated electromagnetic wave (modulated radio wave). This electromagnetic wave travels through air space at the speed of approximately 3×10^8 m/s. When the radio wave strikes an antenna, a modulated high-frequency signal similar to the signal in the transmitting antenna is created in it. A detector stage in the receiver extracts the carrier frequency from the audio signal information. An audio amplifier in the receiver amplifies the audio signal and sends it to a speaker that converts the audio signal into the sound waves. A modulated radio wave is shown as follows:

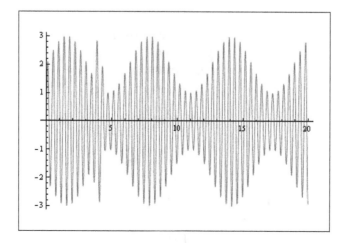

The modulation of the radio wave is called **Amplitude Modulation** (**AM**).

AM was the first method of superimposing sound on a radio signal and is still widely used today. Commercial and public AM broadcasting is authorized in the medium wave band (535-1,705 kHz) worldwide and also in parts of the long wave and short wave bands.

The Arduino-SNAP circuit AM radio block diagram

The Arduino-SNAP circuit AM radio is basically an IR (wireless) controlled receiver. An ordinary IR handheld remote will turn the SNAP circuit AM radio *ON* or *OFF*. The littleBits remote trigger and latch electronic modules detect and send a logic control signal to the Arduino Uno. The Arduino Uno turns the SNAP circuit AM radio *ON* or *OFF* using a small transistor relay driver. In addition to operating the radio, an LCD provides the status of the receiver by displaying *Radio OFF* or *Radio ON* messages. A concept drawing of an IR remote-operated AM radio is shown as follows:

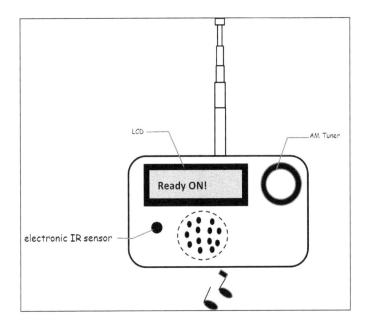

The Arduino-SNAP circuit AM radio block diagram is an engineering development tool used to convey complete product designs using graphics. The block diagram also helps when planning the breadboard for prototyping and testing of the Arduino-SNAP circuit AM radio touch in a maker's workshop or laboratory bench. The final observation of the Arduino-SNAP circuit block diagram is the basic computer convention of having the inputs on the left-hand side, the processor is located in the middle, and the outputs are placed on the right-hand side of the design layout. As shown, the littleBits remote trigger electronic module is on the left-hand side; the Arduino is located in the middle; and the SNAP circuit AM radio and LCD are shown to the right of the block diagram. The SNAP circuit AM radio will turn *ON* or *OFF* based on the littleBits remote trigger detecting an IR signal from a handheld remote. The Arduino-SNAP circuit AM radio block diagram is shown as follows:

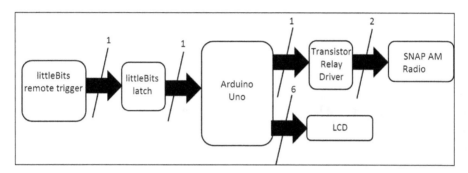

[A simple receiver requiring no power source but power from electromagnetic waves is called a **crystal radio**.]

The littleBits latch electronic module

The littleBits latch electronic module is a digital circuit that provides a bistable signal for allowing the SNAP AM radio to stay *ON* or *OFF* after releasing the button on an ordinary IR handheld remote. The key digital IC component behind the littleBits latch electronic module is a 74HC74 D flip-flop. The 74HC74 D flip-flop circuit toggles its output on receiving a positive edge signal from the littleBits remote trigger module. With every button press from an ordinary IR handheld remote, the SNAP AM radio will turn *ON* or *OFF* based on the 74HC74 D flip-flop toggling operation.

The littleBits latch electronic module and circuit schematic diagram are shown as follows:

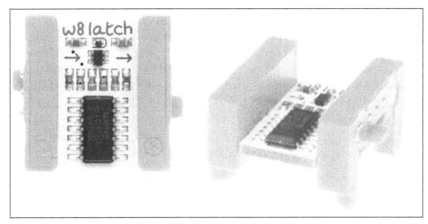

The littleBits latch electronic module

 Flip-flop and latch are words used interchangeably in digital and industrial electronics design and development work.

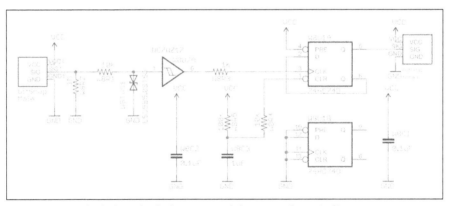

The littleBits latch electronic module circuit schematic diagram

The UTC 7642 linear IC one-chip AM radio circuit

The traditional radio has changed considerably with the use of a single IC capable of receiving AM signals. The IC packaged inside of the SNAP circuit AM radio module is a UTC 7642 linear circuit. The UTC 7642 is a 3-pin IC capable of amplifying the AM radio frequency band of 535-1,705 kHz. The UTC 7642 IC consists of several transistor amplifiers that boost the low AM radio frequencies received from the antenna coil. The SNAP circuit AM radio module, T092 IC component, and the device's internal circuit schematic diagram are shown as follows:

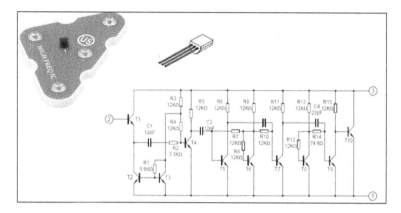

Also, the additional reference material, the block diagram and complete circuit schematic diagram of the SNAP circuit AM radio are provided as follows:

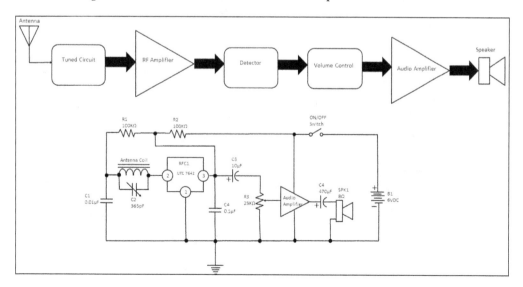

In designing **Printed Circuit Boards** (**PCB**), the T092 IC is a typical footprint used for 3-leaded electronic components.

Building the IR remote trigger circuit

The Arduino-SNAP circuit AM radio consists of several independent circuits that integrate (wire) together to make one awesome electronic device. Therefore, the prototype build will start with the wiring of the littleBits remote trigger to the latch electronic module. Actually, the wiring is more attaching of the two electronic modules by way of magnetic bit snaps. Also, included in this IR remote trigger circuit build is the proto module. The proto module provides the hardwired control signal that the Arduino Uno uses to detect the logic level swings created by the littleBits latch circuit. The complete IR remote trigger interface circuit wiring diagram is shown as follows:

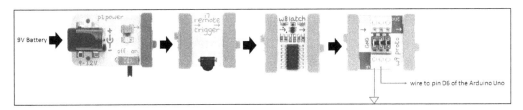

To test the IR remote trigger interface, attach the red test lead of a DC voltmeter to the pin D6 wire on the proto module. The DC voltmeter's black test lead is attached to the proto module's ground wire. With the DC voltmeter set to the proper voltage scale, pressing buttons on the IR handheld remote should toggle the voltage reading between 0 and 5V DC.

Building the Arduino-SNAP circuit interface

With the IR remote trigger interface built and working correctly, the Arduino-SNAP circuit interface is the next building block to construct. As discussed in the *Arduino-SNAP circuit AM radio block diagram* section of this chapter, this electronic circuit interface has two output devices it operates: the LCD and the transistor relay driver. When the Arduino Uno receives a control signal from the IR remote trigger circuit, the LCD will display *Radio ON!* or *Radio OFF* messages and appropriately turns the SNAP circuit AM radio *ON* or *OFF*. The proto module provides the hardwired interface between the IR remote trigger circuit and the Arduino-SNAP circuit AM radio. To build the Arduino-SNAP circuit interface, use the following wiring diagram. Also, the circuit schematic diagram is provided as an additional wiring reference source:

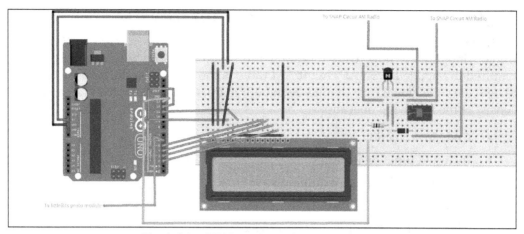

The Arduino-SNAP circuit interface wiring diagram

The following is the circuit schematic diagram provided as an additional wiring reference source:

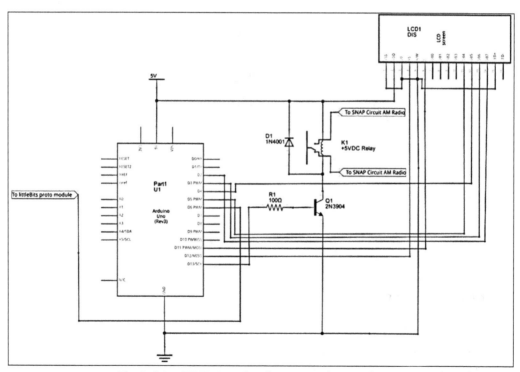

The Arduino-SNAP circuit interface circuit schematic diagram

Testing the Arduino-SNAP circuit interface

With the Arduino-SNAP circuit built on the breadboard and wired to the IR remote trigger circuit, we're now ready to test the mini controller. The IR remote trigger code required to operate the SNAP circuit AM radio is shown later in the text. As you review the code, you'll note that it's been built in operational sections. The sections consist of the following operations:

- Code introduction
- Included function libraries
- Initializing/defining Arduino pins
- Defining Arduino pins as I/O
- Checking the IR remote trigger pin status
- Operational loop (if...else statements)

This method of partitioning the code will help when software debugging is necessary:

```
/*
  IR remoter trigger control

  Turns ON and Off a light emitting diode(LED) connected to
  digital
  pin 13 and the SNAP circuit AM Radio when pressing a button on
  an ordinary IR handheld remote.

  The circuit:
* Transistor Relay driver attached to pin 13
* LED attached from pin 13 to ground
* proto module attached to pin 6
*

* Note: on most Arduinos there is already an LED on the board
  attached to pin 13.

  created April 26 2010
  by Don Wilcher


*/
// include the lcd library code
#include <LiquidCrystal.h>

//initialize the lcd library with the numbers of the interface
pins
```

```
LiquidCrystal lcd (12,11,5,4,3,2);
// constants won't change. They're used here to
// set pin numbers:
const int IRPin = 6;      // the number of the IR remote trigger
pin
const int TrRelayPin =  13;      // the number of the Transistor
Relay driver Pin

// variables will change:
int IRState = 0;          // variable for reading the IR remote
trigger status

void setup() {
  // setup the lcd's number of columns and rows:
  lcd.begin(16,2);

  // initialize the Transistor Relay driver pin as an output:
  pinMode(TrRelayPin, OUTPUT);
  // initialize the IR remote trigger pin as an input:
  pinMode(IRPin, INPUT);

}

void loop() {
  // read the state of the IR remote trigger value:
  IRState = digitalRead(IRPin);

  // check if the IR remote pushbutton is pressed.
  // if it is, the IRState is HIGH:
  if (IRState == HIGH) {
    // turn Transistor Relay Driver on:
    digitalWrite(TrRelayPin, HIGH);
    lcd.setCursor(0,0);
    lcd.print("Radio ON!");

  }
  else {
    // turn Transistor Relay driver off:
    digitalWrite(TrRelayPin, LOW);
    lcd.setCursor(0,0);
    lcd.print("Radio OFF");
  }
}
```

There is an Easter egg in the code. Can you find it?

Hint: it lights up when an IR handheld remote button is pressed. You'll find the answer later in the chapter.

With the code uploaded to the Arduino Uno, the transistor relay driver's contact should turn on with every button pressed on an IR handheld remote. The LCD will display the appropriate operation message (*Radio ON!* / *Radio OFF*) as well. If the transistor relay driver or LCD is not working correctly, correct any wiring and software errors present and repeat the test again. Congratulations on completing the IR remote trigger — Arduino-SNAP circuit interface! The final step for this project is to build the SNAP circuit AM radio and wire the electronic device to the IR remote interface.

Building the SNAP circuit AM radio

The final circuit block for the project is SNAP circuit AM radio. As discussed in the *UTC 7642 linear IC one-chip AM radio circuit* of this chapter, the small 3-pin integrated circuit is the main component of the receiver device. In addition to having a one-radio chip solution, the SNAP circuits make the task of building the receiver easy and fun. The *Parts list* shown earlier in the chapter provided the name of the complete SNAP circuit kit, which includes the electronic modules to build the AM radio. By placing the electronic modules on a plastic board and snapping the components together, the AM radio can be built in several minutes. The complete electronic SNAP circuit AM radio is shown here:

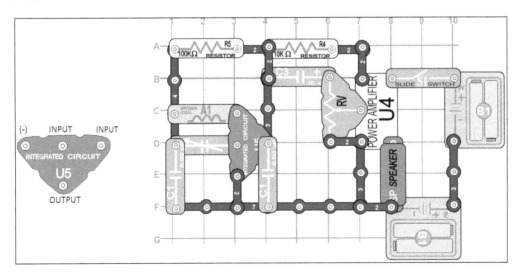

For additional information on SNAP circuits and accessories, refer to the following URL address: http://www.snapcircuits.net/.

After building the circuit, slide the power switch *ON* and adjust the tuner (variable capacitor) to a radio station. Adjust the volume control for the best audio output sound. If the circuit is not working properly, correct the electrical SNAP connections. Also, replace the batteries if they are weak or old. Once the circuit is working correctly, remove the slide switch and replace them with red and black SNAP wires. These wires will be attached to the relay contacts of the Arduino-SNAP circuit interface. The next, and final stage, of this wireless remote control project is the subcircuit integration.

Subcircuit integration

Now that all the subcircuits are working correctly, the final stage of this wirelessly remote control project is to wire them together. Each subcircuit diagram presented in this chapter provides wiring attachment notes to help with the integration of the subcircuits. A couple of tips to help connect these circuits are listed as follows:

- Assure all power sources are working properly (batteries, USB-powered cables, and external DC power supplies)
- Assure all electrical grounds are properly connected together
- Assure all jumper wires do not have exposed conductors

Using the subcircuit integration tips listed, the complete Arduino-SNAP circuit AM radio is presented in the following screenshot:

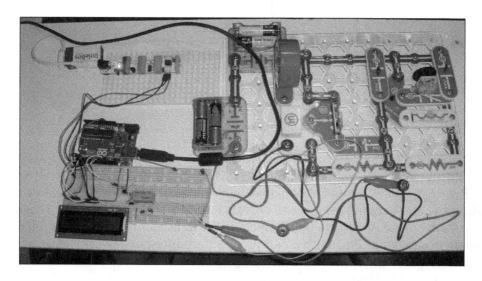

Test the final product by pressing a button on an ordinary IR handheld remote and observing the LCD and radio operating correctly. Awesome work and congratulations on building the Arduino-SNAP circuit AM radio! Your technical knowledge in prototyping complex circuits of various construction formats have increased tremendously. Now you are ready or a DIY challenge that will put your new technical knowledge to the test. Just to prepare you for the challenge, I'll provide a small clue to think on: the littleBits IR remote trigger circuit will be replaced with Bluetooth Smart technology.

The Easter egg is the on-board LED.

DIY challenge – RedBearLab BLE control

If you haven't guessed what the DIY challenge is from the provided clue, it's building a BLE wireless controller. The idea behind this DIY challenge is to replace the littleBits IR remote trigger circuit interface using the RedBearLab BLE shield. The BLE shield makes the product size smaller, and it can be operated by using an Apple iOS or Android smartphone. A Bluetooth-enabled mobile device, such as a tablet, can also be used as the **User Interface** (**UI**) to operate the SNAP circuit AM radio as well. The design parameters for the DIY challenge are based on the block diagram shown as follows:

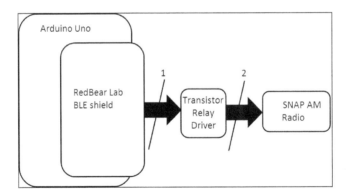

The RedBearLab BLE shield will provide the control signal to operate the transistor relay driver circuit. The contacts of the relay when closed will turn the SNAP circuit AM *Radio ON*. Again, a Bluetooth-enabled tablet or smartphone will provide a UI-based wireless control interface to operate the SNAP circuit AM radio. To aid in the wiring of the RedBearLab BLE shield to the transistor relay circuit, a reference diagram of a previous Bluetooth Smart project, introduced in *Chapter 7, Bluetooth Low Energy Controller,* is included.

The reference wiring diagram is shown in the following image:

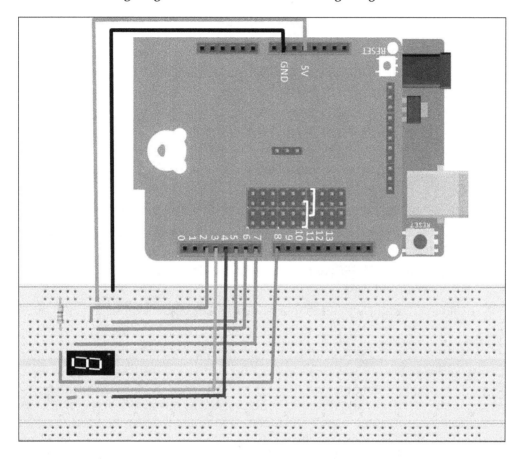

The trick to using this wiring diagram to design and build the new BLE controller is to remove all of the wires and electronic components from the breadboard. The designated pin from the Arduino-SNAP circuit AM radio used to control the transistor relay driver is D13. Therefore, a wire from D13 of the RedBearLab BLE shield should connect to the 470 Ω base resistor of the transistor driver circuit. The operational setup of RedBearLab shield is to upload the BLE controller code to the Arduino Uno. To aid in uploading the code to the Arduino Uno, refer to *Chapter 6, A Simple Chat Device with LCD* for the procedure. Once the BLE controller code has been uploaded to the Arduino Uno, the RedBearLab mobile app needs to be set up properly as well. Consult *Chapter 7, Bluetooth Low Energy Controller* for details on proper mobile app setup. Remember that D13 is the designated digital pin to control with the mobile app. Document all new circuit design changes and procedures in your lab notebook as well. Good luck and have fun with this DIY challenge!

Summary

In this chapter, an Arduino-SNAP circuit AM radio prototype was built:

- A discussion on radio communication basics was presented

- Modulation is a radio wave processing technique in which two distinct frequencies are superimposed to produce one unique signal, and it was presented in this chapter as well

- The UTC 7642 linear amplifier IC was introduced along with its internal circuit schematic diagram and device pictorial diagram

- This IC, with the ability to boost high-frequency signals, is the heart of the SNAP circuit AM radio

- The IR remote trigger circuit was discussed using a picture diagram.

- The IR remote sensor provides logic level voltages that allow the Arduino to operate the LCD and SNAP circuit AM radio

- To aid in the *ON/OFF* control of the SNAP circuit AM radio, a littleBits latch was introduced

- A latch is a digital circuit that provides a stable logic level voltage (0V or +5V) based on the device being triggered by a positive edge input signal

To test the IR remote trigger circuit with the SNAP circuit interface, the operational code was uploaded to the Arduino. The various sections of the code were discussed along with a simple testing method to check the Arduino-SNAP circuit interface operation. The SNAP circuit AM radio build phase was presented next in the project discussion. By applying 6V DC to the AM radio adjusting the tuner (variable capacitor) will allow stations to be heard through the speaker. The audio amplitude can be adjusted with the volume control (variable resistor / potentiometer). The final stage of the chapter project consists of integrating or wiring all the subcircuits together. Several integration tips were provided in the discussion as well.

Finally, a DIY challenge in this chapter was presented. The DIY challenge consists of replacing the littleBits IR remote trigger circuit with a Bluetooth Smart design. The RedBearLab shield block diagram was provided as a design reference tool. Also, the RedBearLab seven segment LED controller was provided as a design reference for assisting in wiring the transistor relay driver to pin D13. *Chapter 6, A Simple Chat Device with LCD* and *Chapter 7, Bluetooth Low Energy Controller* were provided as the reference material for code installation and setup of the BLE controller mobile app to operate the SNAP-circuit AM radio.

In the next and final chapter, *Chapter 10, Arduino Scrolling Marquee* will be discussed. Physical computing techniques will be explored to control the speed and contrast of a scrolling marquee display. A discussion of how **organic light-emitting diode's (OLED's)** work will be presented in this final chapter as well. Also, an exploration of how to send messages using the Arduino serial monitor will be provided. The following items to be discussed in the next chapter are:

- Building a simple LCD marquee
- The theory and operation of OLEDs
- Operating a scrolling OLED marquee display with physical computing techniques
- Sending text messages from an Arduino serial monitor to an OLED marquee display

As usual, mini lab-testing procedures, Arduino code explanations, and the final project assembly will be discussed in the next chapter.

10

Arduino Scrolling Marquee

A **marquee display** is an electronic customizable form of signage that shows individual movable letters, numbers, or words. They are used to display critical messages or information from a distance. Marquee displays are usually mounted several feet from the ground, so they become visible. Marquee displays are used by movie theaters to list current film rosters and industrial companies to provide manufacturing processing information. Public transportation, such as buses and trains, uses scrolling marquees to display the route destination and vehicle numbers. Also, warning signs and traffic information use scrolling marquees. Providing manufacturing processing information using marquee displays requires an industrial controller capable of reading electrical sensor signals and displaying them on the electronic signage device.

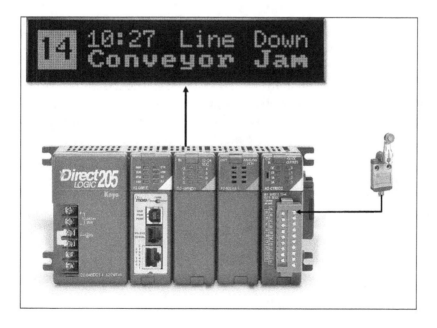

To make the letters, numbers, and characters scroll across the marquee, each pixel is addressable or controlled using binary bit values. The industrial controller provides the binary bit sequences to allow messages to be visible and scroll across the marquee display. Typically, LEDs and **Liquid Crystal Display (LCDs)** are used in marquee displays to show discrete letters, numbers, and characters. An important concern regarding the use of these optoelectronic devices is the heat generated and viewing distance. Discrete LEDs get warm with extended use. Although LCDs operate at cooler temperatures, they require internal backlighting so that they are seen at nighttime. Usually, the backlighting element is an LED, which could contribute to small increases in temperature to the LCD. To avoid temperature and visual concerns, **organic light-emitting diodes (OLEDs)** are now being used in marquee displays.

In this chapter, you will learn how to build a scrolling marquee using an Arduino Uno and OLED LCD. You will learn how to control the scrolling direction of the message displayed on the OLED LCD using an electronic IR sensor. The littleBits IR remote sensor and an ordinary IR handheld remote will be used to control the direction of the scrolling text on the OLED LCD. Also, the chapter's DIY challenge will allow you to learn how to send messages to the OLED LCD using the Arduino's IDE serial monitor. The parts required to build the scrolling marquee device are listed as follows:

Parts list

The following is the list of parts required to build a scrolling marquee device:

- Arduino Uno (R3 version) or equivalent (one unit)
- littleBits remote trigger (one unit)
- littleBits latch (one unit)
- littleBits power (one unit)

- littleBits proto module (one unit)
- littleBits mounting board (one unit)
- 16x4 OLED LCD New Haven Display: NHD-0420DZW-AY5 or equivalent (one unit)
- 16x2 LCD JHD 162A or equivalent (one unit)

The LCD and OLED basics

The (**LCD**) is traditionally used to show data, graphics, or both, on electrified glass plates. Typical LCD parts consist of a controller and a glass substrate material. LCD segments mounted on the glass substrate material are operated by an electronic controller that consists of a microprocessor or microcontroller. The LCD's crystal segments are placed between glass plates with electrodes. In order for the crystal segments to form letters or characters, a small AC RMS voltage is needed. The AC RMS voltage turns on the crystal segments. The electronic device responsible for managing this electrified voltage for the crystal segments is the LCD controller. To assure the timing sequences to turn the crystal segments on or off are correct, a microprocessor or microcontroller is used inside of the LCD controller. Typical LCD controllers used to operate the crystal segments are HD44780 and KS0066 devices. A typical LCD and its controller, identified by U2, are shown as follows:

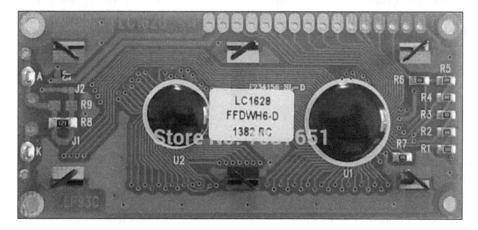

 RMS is the abbreviation for **Root Mean Square**, a continuously varying signal. RMS is usually associated with AC voltage and current in electrical circuits.

A typical block diagram of an LCD with its controller is shown as follows:

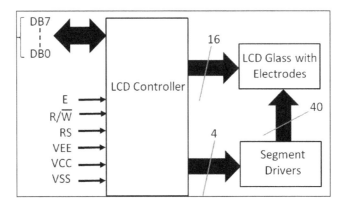

An OLED works in a similar way to general purpose diodes and LEDs. Instead of using layers of *n-type* and *p-type* doping semiconductor materials, OLEDs use organic molecules to produce their *n-type* and *p-type* semiconductor layers. A typical OLED is made up of six different layers. The OLED has layers of protective glass and plastic placed on the top and bottom of the optoelectronic device. The top layer is known as **seal**, and the bottom is called **substrate**. In between those layers, there is a negative terminal called **cathode** and a positive connection point known as **anode**. In between the cathode and anode terminals, there are two layers made from organic molecules. The technical name for these organic molecule layers is called the **emissive layer**. Light is produced by the emissive layer when the cathode and anode terminals are wired correctly to a voltage source. A sectional view of an OLED is shown in the following image:

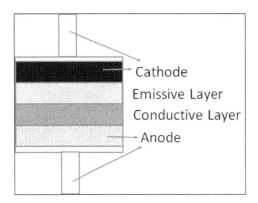

 Doping is the process of adding impurities to pure semiconductor materials.

There are two other varieties of OLEDs being used in visual display electronics called SMOLEDs and AMOLEDs. **Small molecule organic light-emitting diodes (SMOLEDs)** work like traditional OLEDs. The small organic molecules allow the OLED size to be reduced. Therefore, they can be used in small devices, such as wristband fitness trackers, which require information to be displayed. Another application of SMOLEDs is in tactile switches. For example, pushing the SMOLED-based button on a tactile switch can display the words *ON* or *OFF*. Mobile phones and small display for wearable devices (FitBit) also use SMOLEDs to display information such as the time or date.

AMOLEDs or **active matrix organic light-emitting diodes** is a new technology being used in mobile phones and televisions. AMOLEDs use organic materials to help the diode with luminescence. The active matrix refers to the technology behind the addressing of pixels of the AMOLED display. An embedded microprocessor provides the addressing scheme of activating the AMOLED pixels.

Like the ordinary LCD, an OLED display has a controller to manage the electrified voltage for the organic material segments. A typical OLED with a display controller block diagram is shown as follows. The 6800 **Microprocessor Unit (MPU)** operates the OLED LCD glass by generating the proper software timing sequences to control the organic-based segments. Now, we will explore how to make an Arduino scrolling marquee using software and off-the-shelf electronics:

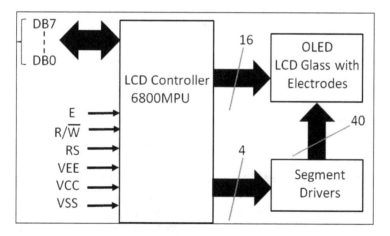

 The 6800 is an 8-bit microprocessor designed and first manufactured by Motorola in 1974.

The Arduino scrolling marquee block diagram

The Arduino scrolling marquee uses an OLED LCD to display moving messages. The OLED LCD marquee is designed to scroll messages either left or right. Providing a physical computing interface allows the scrolling marquee messages to be controlled by an external trigger such as light, sound, or pressure. As discussed in the previous chapter, an ordinary IR handheld remote can change the direction of the scrolling message by pressing any button on the unit. The littleBits remote trigger and latch electronic modules detect and send a logic control signal to the Arduino Uno. The Arduino Uno controls the direction of the marquee's scrolling messages based on the presence or absence of an IR signal from the handheld remote. A concept drawing of an Arduino scrolling marquee block diagram is shown as follows:

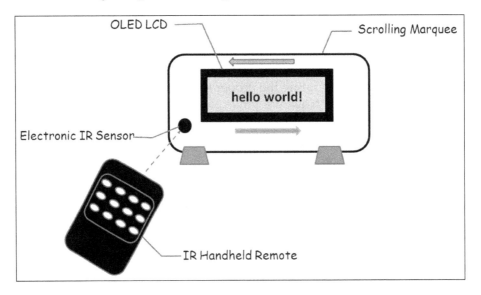

The Arduino scrolling marquee block diagram is an engineering development tool used to convey complete product designs using graphics. The block diagram also makes it easy in planning the breadboard for prototyping and testing of the Arduino scrolling marquee in a Maker's workshop or laboratory bench. A final observation of the Arduino scrolling circuit block diagram is the basic computer convention of the inputs to the left, the processor located in the middle, and the outputs placed on the right-hand side of the design layout. As shown, the littleBits remote trigger electronic module is on the left-hand side; the Arduino is located in the middle; and the OLED LCD is shown to the right of the block diagram. The scrolling marquee message changes direction based on the littleBits remote trigger detecting an IR signal from a handheld remote.

The Arduino scrolling marquee block diagram is shown as follows:

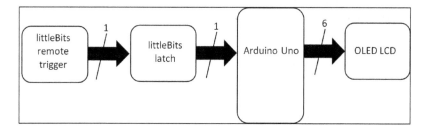

Wiring the OLED LCD

The OLED LCD is quite easy to wire to Arduino. Basically, there are six control lines to wire to the Arduino Uno. The control lines allow the Arduino to enable the LCD controller to turn on the segments to make letters and characters for the scrolling message. Improper wiring of these six control lines will not display the desired message correctly. Also, the OLED LCD requires a +5V DC supply to operate properly. The Arduino Uno has a +5V DC power supply along with three electrical grounds. Wiring the +5V DC to pin 2 (VDD) and ground to pin 1 (VSS) will ensure that the OLED LCD is powered up correctly. To help connect the Arduino Uno to the OLED LCD correctly a wiring diagram is provided in the following image:

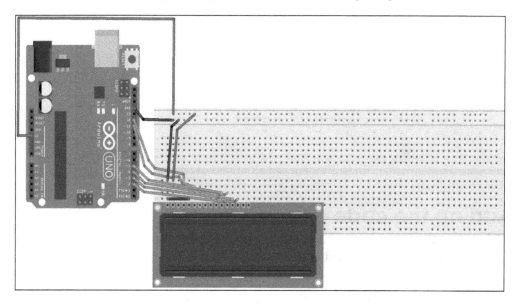

 VSS and **VDD** are the abbreviations for the **source supply** and **drain supply** voltages.

Also, the circuit schematic diagram is provided as another wiring reference resource to aid in building the scrolling marquee device. The reference circuit schematic diagram is shown as follows:

> Before applying to the scrolling marquee circuit, recheck your wiring using both the electrical diagrams.

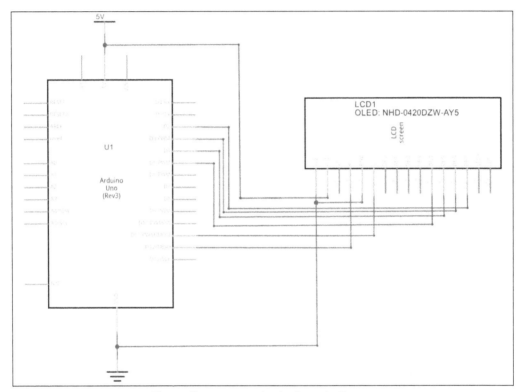

The scrolling marquee circuit schematic diagram

Adding the code

With the scrolling marquee circuit built, it's time to bring the device to life. Enter the following code into the IDE text editor and upload it to the Arduino Uno:

```
/*
OLED _Marquee LiquidCrystal Library - Autoscroll

Demonstrates the use a 16x4  OLED LCD.
```

```
This sketch demonstrates the use of the autoscroll()
and no Autoscroll() functions to make new text scroll or not.

The circuit:
* LCD RS(pin 4) to Arduino digital pin 12
* LCD Enable(pin 6) to Arduino digital pin 11
* LCD D4 (pin 11)to Arduino digital pin 5
* LCD D5 (pin 12) to Arduino digital pin 4
* LCD D6 (pin 13) to Arduino digital pin 3
* LCD D7 (pin 14) to Arduino digital pin 2
* LCD R/W (pin 5) to ground

*/

// include the library code:
#include <LiquidCrystal.h>

// initialize the library with the numbers of the interface pins
LiquidCrystal lcd(12, 11, 5, 4, 3, 2);

void setup() {
  // set up the LCD's number of columns and rows:
  lcd.begin(16, 4);
}

void loop() {
  // set the cursor to (0,0):
  lcd.setCursor(0, 0);
  // print from 0 to 9:
  for (int thisChar = 0; thisChar < 10; thisChar++) {
    lcd.print(thisChar);
    delay(500);
  }

  // set the cursor to (16,1):
  lcd.setCursor(16, 1);
  // set the display to automatically scroll:
  lcd.autoscroll();
  // print from 0 to 9:
  for (int thisChar = 0; thisChar < 10; thisChar++) {
    lcd.print(thisChar);
    delay(50);
  }
```

```
  // turn off automatic scrolling
  lcd.noAutoscroll();

  // clear screen for the next loop:
  lcd.clear();
}
```

The OLED LCD will show a sequential counter with a series of count values (0, 1, 2, 3, 4, 5, 6, 7, 8, and 9). It takes the autoscroll feature of this code 50 **milliseconds** (**ms**) for each number to print on the OLED LCD. Also, the autoscroll feature of the code scrolls the numbers on the OLED LCD from right to left. The numbers also wrap when starting from the second line on the OLED LCD. The lines of code responsible for printing the count values every millisecond on the OLED LCD is shown as follows:

```
  // set the display to automatically scroll:
  lcd.autoscroll();
  // print from 0 to 9:
  for (int thisChar = 0; thisChar < 10; thisChar++) {
    lcd.print(thisChar);
    delay(50);
```

In addition, the autoscroll code initially prints the count values at (0, 0) of the OLED LCD and then displays them on the second line at location (16, 1) of the optoelectronic device. In all cases, as the second line displays numbers, the first line scrolls another set of the count values starting at location (16, 1). The autoscroll code is a well-choreographed visual effect seen on the OLED LCD. The lines of code used to display the count values at (16, 1) is listed as follows:

```
  // set the cursor to (16,1):
  lcd.setCursor(16, 1);
  // set the display to automatically scroll:
  lcd.autoscroll();
  // print from 0 to 9:
  for (int thisChar = 0; thisChar < 10; thisChar++) {
    lcd.print(thisChar);
    delay(50);
```

Try experimenting with the `delay` values to allow the count values to move across the OLED LCD at different scroll rates. Also, change the `lcd.setCursor` coordinate values to set different starting print locations for the count values on the OLED LCD as well. Remember to record your new discoveries in a lab notebook! Congratulations on building your OLED LCD scrolling marquee!

Building an IR-controlled scrolling marquee

The IR-controlled scrolling marquee is a remix of the project of *Chapter 9, Arduino-SNAP Circuit AM Radio*, where the infrared interface circuit will operate a scrolling marquee. As shown in the block diagram, the IR interface circuit provides the input control signal to allow directional operation of the scrolling marquee. The OLED LCD is wired to six control lines of the Arduino Uno. These six control lines provide binary information to operate the OLED LCD controller to display the scrolling messages on the optoelectronic device.

Pressing a button on an ordinary IR handheld remote will command the OLED LCD to scroll the message *hello world!* in opposite directions. Besides the littleBits remote trigger being an important component for this electronic device, the latch electronic module allows the messages to scroll in opposite directions across the OLED LCD without pressing the IR handheld remote button continuously. The reaction time for switching directions is quite responsive (150 ms), which makes the control element of the marquee's operation real time.

To test the IR interface circuit prior to wiring it to the Arduino Uno, take a digital or analog voltmeter's red test lead and attach it to the wire that connects the littleBits proto module to pin D6 of the Arduino Uno. Attach the black test lead of the voltmeter to the ground of the interface circuit. Pressing the button on an IR handheld remote will display either 0V or +4.99V on the voltmeter. If a digital or analog voltmeter is not available, wiring a series current-limiting resistor with an LED to the littleBits proto module will provide a simple visual indicator that the IR interface circuit is working properly.

As discussed in the previous section of this chapter, the Arduino Uno has a +5V DC power supply along with three electrical grounds. Wiring the +5V DC to pin 2 (VDD) and ground to pin 1 (VSS) will ensure that the OLED LCD is powered up correctly. To help connect the Arduino Uno to the OLED LCD and the IR interface circuit correctly, the wiring diagram to build the IR-controlled scrolling marquee is shown as follows:

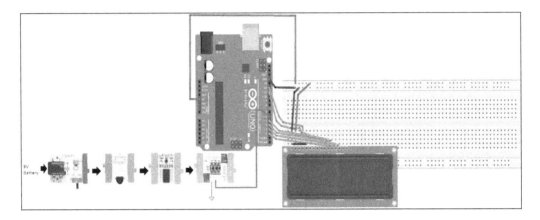

 The proto module allows other littleBits input electronic modules to control the scrolling marquee quite easily.

Also, the circuit schematic diagram is provided as another wiring reference resource to aid in building the controlled scrolling marquee device. The reference circuit schematic diagram is shown as follows:

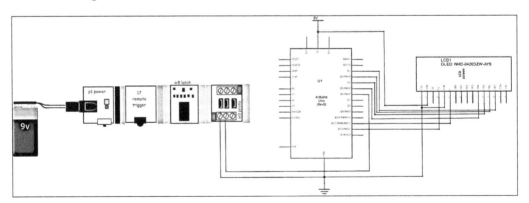

Adding the IR code

With the IR interface circuit built, tested, and wired to the Arduino Uno, the IR code can be developed. The code used to allow an ordinary IR handheld remote to control the scrolling direction on an OLED LCD is shown as follows:

```
/*
This sketch prints "Hello World!" to the LCD and uses the
scrollDisplayLeft() and scrollDisplayRight() methods to scroll
the text.

The circuit:
* LCD RS(pin 4) to Arduino digital pin 12
* LCD Enable(pin 6) to Arduino digital pin 11
* LCD D4 (pin 11)to Arduino digital pin 5
* LCD D5 (pin 12) to Arduino digital pin 4
* LCD D6 (pin 13) to Arduino digital pin 3
* LCD D7 (pin 14) to Arduino digital pin 2
* LCD R/W (pin 5) to ground

*/

// include the library code:
#include <LiquidCrystal.h>

// constants won't change. They're used here to
// set pin numbers:
const int triggerPin = 6;      // the number of the trigger device
pin
const int ledPin =  13;      // the number of the LED pin

// variables will change:
int triggerState = 0;          // variable for reading the trigger
device status

// initialize the library with the numbers of the interface pins
LiquidCrystal lcd(12, 11, 5, 4, 3, 2);

void setup() {
   // initialize the LED pin as an output:
  pinMode(ledPin, OUTPUT);
   // initialize the pushbutton pin as an input:
  pinMode(triggerPin, INPUT);
```

```
  // set up the LCD's number of columns and rows:
  lcd.begin(16, 4);
  // Print a message to the LCD.
  //lcd.setCursor(0,3);
  lcd.print("hello, world!");
  delay(20);
}


void loop() {
  // read the state of the pushbutton value:
  triggerState = digitalRead(triggerPin);

  // check if the trigger device is active.
  // if it is, the triggerState is HIGH:
  if (triggerState == HIGH) {
    // turn LED on:
    digitalWrite(ledPin, HIGH);
    // scroll 13 positions (string length) to the left
  // to move it offscreen left:
    for (int positionCounter = 0; positionCounter < 13;
      positionCounter++) {
      // scroll one position left:
      lcd.scrollDisplayLeft();
      // wait a bit:
      delay(150);
    }
  }
  else {
    // turn LED off:
    digitalWrite(ledPin, LOW);
    // scroll 29 positions (string length + display length) to the
    right
    // to move it offscreen right:
    for (int positionCounter = 0; positionCounter < 29;
      positionCounter++) {
      // scroll one position right:
      lcd.scrollDisplayRight();
```

```
    // wait a bit:
    delay(150);
  }
 }
}
```

The IR operation is quite easy to understand from the comments inserted into the code. The IR code provides a convenient way to control the message scrolling direction on the OLED LCD using any household infrared remote. As always, experiment with the message scrolling rate by way of the Arduino delay instruction. Also, change the lcd.print instruction to customize the message. Record new changes and discoveries in your lab notebook and a big congratulations for building the IR-controlled scrolling marquee! Now, your OLED LCD skills will be put to the challenge by building a serial monitor LCD marquee.

A DIY serial monitor LCD marquee

Besides being able to use the serial monitor to view variable values during a code debugging session, it can also send messages to an LCD. The basic concept drawing for the DIY serial monitor LCD marquee is shown as follows:

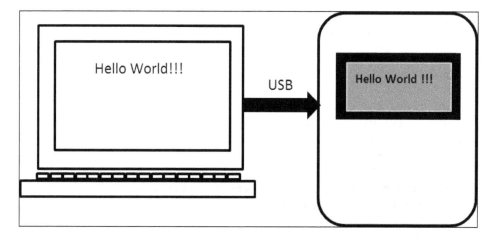

This DIY challenge is based on text messaging concepts discussed in *Chapter 6, A Simple Chat Device with LCD*. The idea behind this challenge is to test and compare the operation and esthetics of an ordinary LCD to an OLED display device. The internal structure of the concept drawing can be viewed as a small LCD/OLED display marquee connected to Arduino Uno. Sending messages to the marquee is done through a standard USB cable connected to the Arduino Uno as shown in the following block diagram:

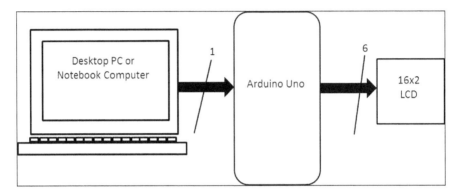

Although, the block diagram shows the Arduino Uno wired to a 16x2 LCD, an OLED LCD can easily be used in the serial monitor marquee device. The code to make the serial monitor marquee work is shown as follows:

```
/*
  LiquidCrystal Library - Serial Input

  Demonstrates the use a 16x2 LCD display.  The LiquidCrystal
  library works with all LCD displays that are compatible with the
  Hitachi HD44780 driver. There are many of them out there, and you
  can usually tell them by the 16-pin interface.

  This sketch displays text sent over the serial port
  (e.g. from the Serial Monitor) on an attached LCD.

  The circuit:
  * LCD RS pin to digital pin 12
  * LCD Enable pin to digital pin 11
  * LCD D4 pin to digital pin 5
  * LCD D5 pin to digital pin 4
  * LCD D6 pin to digital pin 3
  * LCD D7 pin to digital pin 2
```

```
 * LCD R/W pin to ground
 * 10K resistor:
 * ends to +5V and ground
 * wiper to LCD VO pin (pin 3)

*/

// include the library code:
#include <LiquidCrystal.h>
// Arduino digital pin number assigned to the littleBits number
modules
int littleBits_Number = 6;



// initialize the library with the numbers of the interface pins
LiquidCrystal lcd(12, 11, 5, 4, 3, 2);

void setup() {
  // set up the LCD's number of columns and rows:
  lcd.begin(16, 2);
  // initialize the serial communications:
  Serial.begin(9600);
  //littleBits Number module digital pin configured as an output
  pinMode(littleBits_Number, OUTPUT);
  //digitalWrite(littleBits_Number, LOW);
}

void loop()
{
  // when characters arrive over the serial port...
  if (Serial.available()) {

    // wait a bit for the entire message to arrive
    delay(100);
    // clear the screen
    lcd.clear();
    // read all the available characters
    while (Serial.available() > 0) {
      // display each character to the LCD
      lcd.write(Serial.read());
      // turn on littleBits Number module during text message
transmission
```

```
      digitalWrite(littleBits_Number, HIGH);
      // transmit text message and display 99 value on littleBits
number module for 100ms
      delay(100);
      //stop text message transmission and display 00 value on
littleBits number module
      digitalWrite(littleBits_Number, LOW);
    }
  }
}
```

As seen in the code, the littleBits number module is included for an added visual indicator. Modify the code as needed to meet the basic text messaging requirement of this DIY challenge. Again, record all changes and discoveries in your lab notebook and by all means have fun and enjoy this last DIY project of the book!

Summary

In this chapter, an Arduino scrolling marquee prototype was built:

- A discussion on LCD and OLED basics was presented
- An LCD controller is a dedicated microprocessor or microcontroller that operates the crystal segments of the optoelectronic device
- An OLED operates like a general purpose diode and LED
- The only difference between general purpose diodes and LEDs, and OLEDs, is the use of organic materials for the anode and cathode terminals to allow light to be emitted from the device when wired correctly to a DC power supply
- This organic material makes OLEDs more efficient in power consumption and visually appealing than the traditional LCDs

Two scrolling marquee projects were discussed in this chapter. An automatic scrolling marquee was presented where a series of count values (0, 1, 2, 3, 4, 5, 6, 7, 8, and 9) move across the OLED LCD at two starting locations (0, 1) and (16, 1). These two starting locations provide an esthetically pleasing look on the OLED LCD. The IR-controlled scrolling marquee allows the direction of the message to change by using an ordinary IR handheld remote. Pressing a button on the IR handheld remote allows the direction of the scrolling message to switch. The littleBits electronics modules make prototyping the IR interface quick and easy.

Finally, a DIY challenge in this chapter was presented. The DIY challenge consisted of using a serial monitor to send messages to an LCD or OLED display. Information from *Chapter 6, A Simple Chat Device with LCD* on text messaging was referenced to aid in meeting the DIY requirements. The code from *Chapter 6, A Simple Chat Device with LCD* is referenced as well to aid in building the DIY project. Also, with the knowledge obtained in writing Arduino code, building electronic circuits, using a variety of input and output component devices, you now have the skills to create sophisticated devices, such as microrobots, electronic games, mini music players, and wireless remote controls. These devices can easily be built by taking project snippets from this book and adding to your own circuit designs. Always, document your electronic designs and code in a notebook to share with other aspiring Arduino fans! Happy Making!!!

Index

Symbols

SPI communication
  defining 6-10
subscription-based circuit simulation
    package
  URL 182
superheterodyne (superhet) 189

T

talking logic probe
  about 58
  block diagram 56, 57
  building 66-72
  code 72-77
  DecTalk speech synthesizer engine 78
  EMIC 2 TTS module, testing 58-62
Telelocator Alphanumeric Protocol
    (TAP) 120
teleprinter 119
Teletype (TTY) 120
TMRpcm library
  about 14
  installing 14-18
  URL 14

Transistor motor driver 46
Transistor-Transistor Logic (TTL) 56
truth table, programmable DC motor
    controller 42
Truth Table (TT) 55

U

User Interface (UI) 155, 202

V

Vx 78

W

WAV file library
  adding, to Arduino sketch 12, 13
wireless 156
Wx 79

Thank you for buying
Arduino Electronics Blueprints

About Packt Publishing

Packt, pronounced 'packed', published its first book, *Mastering phpMyAdmin for Effective MySQL Management*, in April 2004, and subsequently continued to specialize in publishing highly focused books on specific technologies and solutions.

Our books and publications share the experiences of your fellow IT professionals in adapting and customizing today's systems, applications, and frameworks. Our solution-based books give you the knowledge and power to customize the software and technologies you're using to get the job done. Packt books are more specific and less general than the IT books you have seen in the past. Our unique business model allows us to bring you more focused information, giving you more of what you need to know, and less of what you don't.

Packt is a modern yet unique publishing company that focuses on producing quality, cutting-edge books for communities of developers, administrators, and newbies alike. For more information, please visit our website at www.packtpub.com.

About Packt Open Source

In 2010, Packt launched two new brands, Packt Open Source and Packt Enterprise, in order to continue its focus on specialization. This book is part of the Packt Open Source brand, home to books published on software built around open source licenses, and offering information to anybody from advanced developers to budding web designers. The Open Source brand also runs Packt's Open Source Royalty Scheme, by which Packt gives a royalty to each open source project about whose software a book is sold.

Writing for Packt

We welcome all inquiries from people who are interested in authoring. Book proposals should be sent to author@packtpub.com. If your book idea is still at an early stage and you would like to discuss it first before writing a formal book proposal, then please contact us; one of our commissioning editors will get in touch with you.

We're not just looking for published authors; if you have strong technical skills but no writing experience, our experienced editors can help you develop a writing career, or simply get some additional reward for your expertise.

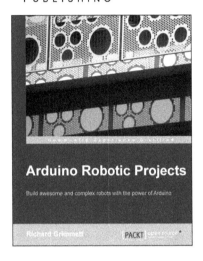

Internet of Things with the Arduino Yún

ISBN: 978-1-78328-800-7   Paperback: 112 pages

Projects to help you build a world of smarter things

1. Learn how to interface various sensors and actuators to the Arduino Yún and send this data in the cloud.

2. Explore the possibilities offered by the Internet of Things by using the Arduino Yún to upload measurements to Google Docs, upload pictures to Dropbox, and send live video streams to YouTube.

3. Learn how to use the Arduino Yún as the brain of a robot that can be completely controlled via Wi-Fi.

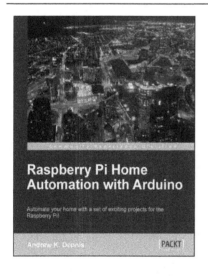

Raspberry Pi Home Automation with Arduino

ISBN: 978-1-84969-586-2   Paperback: 176 pages

Automate your home with a set of exciting projects for the Raspberry Pi!

1. Learn how to dynamically adjust your living environment with detailed step-by-step examples.

2. Discover how you can utilize the combined power of the Raspberry Pi and Arduino for your own projects.

3. Revolutionize the way you interact with your home on a daily basis.

Please check **www.PacktPub.com** for information on our titles

www.ingramcontent.com/pod-product-compliance
Lightning Source LLC
Chambersburg PA
CBHW082116070326
40690CB00049B/3557